Waste management planning and optimisation

Handbook for municipal waste prognosis and sustainability assessment of waste management systems

Emilia den Boer
Jan den Boer
Johannes Jager
[Eds.]

ibidem-Verlag
Stuttgart

European Commission

ENERGY, ENVIRONMENT
AND SUSTAINABLE DEVELOPMENT

RTD-Project "The Use of Life Cycle Assessment Tools for the Development of Integrated Waste Management Strategies for Cities and Regions with Rapid Growing Economies"

Contract Number: EVK4-CT-2002-00087

Project Duration: September 2002 - August 2005

funded by the European Commission's Fifth Framework Programme under the work programme "Energy, Environment and Sustainable Development", Key Action: "The City of Tomorrow and Cultural Heritage".

The "Handbook for municipal waste prognosis and sustainability assessment of waste management systems: Waste management planning and optimisation" is a product of the above mentioned project and is property of project partners.

Compilation and editing:	Emilia den Boer Jan den Boer Gustavo Barreto
Design:	Jan den Boer Gustavo Barreto
Cover design:	Emilia den Boer
Cover image:	Old Landfill of Wrocław (TUD)
Proofreading:	Patrick Stephenson

Information on how to obtain this Handbook, its translated versions (into Greek, Lithuanian, Polish, Slovak and Spanish) and further project Deliverables:

www.lca-iwm.net

Emilia den Boer, Jan den Boer, Johannes Jager (Eds.)

WASTE MANAGEMENT PLANNING AND OPTIMISATION

Handbook for municipal waste prognosis and sustainability assessment of waste management systems

ibidem-Verlag
Stuttgart

Bibliografische Information Der Deutschen Bibliothek

Die Deutsche Bibliothek verzeichnet diese Publikation in der Deutschen Nationalbibliografie; detaillierte bibliografische Daten sind im Internet über <http://dnb.ddb.de> abrufbar.

∞

Gedruckt auf alterungsbeständigem, säurefreien Papier
Printed on acid-free paper

ISBN: 3-89821-519-9

Printed in Germany

Word from the editors

Tarragona, 3rd of October 2002. Kick-off meeting of the LCA-IWM Project. Most of the project partners see most of the project partners for the first time. Start of a three year long cooperation of companies, institutes and municipalities from throughout the European Union and (then still) the Accession Countries. All of us have noticed differences. Different, sometimes even exotic, names, languages and cultures. Different habits, styles and working methods. Different expectations and goals. Different understanding of common words. Different understandings of the Description of Work.

It is like John Travolta told Samuel L. Jackson in Pulp Fiction: “You know, the funniest thing about Europe is... it’s these little differences”. It were these mostly little, sometimes bigger differences, which made the LCA-IWM Project very interesting to work at. All of us, including ourselves as project coordinators, have learned during these three years. Learned a lot about Sustainability Assessment of Municipal Solid Waste Management Systems, but also about people and about where they come from.

At the same time these differences have been difficulties at times as well. It was not always easy to merge different individual opinion to a common approach. Now, when the Project End is nigh, we feel free to say that the difficulties have been overcome. The project objectives have been achieved: this Handbook and both a Waste Prognostic Tool and an Assessment Tool have been created. They are available to municipalities throughout Europe to support them in designing and improving Municipal Waste Management Systems.

We would like to use the opportunity to thank all the project partners for the past three years. Not only for bringing the LCA-IWM Project to a good end, but especially also for enthusiastic participation, open and honest criticism, pleasant cooperation and of course, for the wonderful project meetings in which the local eating and drinking habits were lavishly shared. Apart from the project partners which are listed hereafter we would also like to thank all the people behind the scenes both in our institute and at all partners: the colleagues that have been bothered with questions, the helping hands supporting at project meetings, the secretaries, system administrators and all those that were helping and supporting us. Last, but not least are very grateful to the European Commission for the financial support of the Project, to the involved municipalities for openness and co-operation and to Marina Franke of Procter & Gamble for her support.

We hope everybody involved in the LCA-IWM Project has enjoyed it. The reader we wish a pleasant reading and we truly hope this Project and Handbook will be a positive contribution to the improvement of the Waste Management situation in Europe.

Darmstadt, July 2005.

The LCA-IWM Project Coordinators

Emilia den Boer
Jan den Boer
Johannes Jager

Preface and Acknowledgement

This Handbook reflects a unique expert cooperation across European Union (Austria, Germany, Greece, Lithuania, Luxembourg, Netherlands, Poland, Slovakia, Spain) on the three pillars of Sustainability: environmental protection, economic development and social responsibility. The project`s focus is on sustainable solid waste management and is a result of a multistakeholder cooperation between experts from universities, industry and consulting engineers.

The result of this EU research project "The Use of LCA (Life Cycle Assessment) Tools for the Development of Integrated Waste Management (IWM) in rapid growing economies" are two essential decision support tools, urgently needed for the enlarged European Union:

1. Waste Prognosis Tool - supporting the prediction of future amounts of generated waste for cities and regions with rapid growing economies
2. Waste Management Assessment Tool - supporting the planning and assessment of waste management strategies.

The use of "LCA for introducing IWM" in regions can be calculated via a computer model which allows different scenarios including: storage of waste at households, collection and transport, waste treatment options (incl. incineration, biological treatment, materials recycling) and disposal. The first use of this Waste Management Assessment Tool is provided via five Case Studies in European Municipalities in Greece, Lithuania, Poland, Slovakia and Spain.

Overall this Handbook provides a unique mix of science, knowhow and practical experience. Although this Handbook includes various assumptions and simplifications, it provides an essential tool for municipalities, decision makers and industry. This is a very valuable project providing future perspective on Sustainable Waste Management

Prof. Dr.-Ing. Marina Franke
Manager Sustainable Development
Procter & Gamble Service GmbH
Honorary Professor at Technical University Darmstadt – Germany

Table of contents

LCA-IWM Project Partners

INSTITUTE FOR WATER SUPPLY AND GROUNDWATER PROTECTION, WASTEWATER TECHNOLOGY, WASTE MANAGEMENT, INDUSTRIAL MATERIAL CYCLES & ENVIRONMENTAL PLANNING (INSTITUT WAR)
TECHNISCHE UNIVERSITÄT DARMSTADT (TUD)
Emilia den Boer
Jan den Boer
Jan Berger
Johannes Jager

GRUP AGA (ANALISI I GESTIO AMBIENTAL)
UNIVERSITAT ROVIRA I VIRGILI (URV-AGA)
Montse Meneses
Julio Rodrigo
Francesc Castells

INSTITUTE OF WASTE MANAGEMENT, DEPARTMENT OF WATER, ATMOSPHERE AND ENVIRONMENT
BOKU - UNIVERSITY OF NATURAL RESOURCES AND APPLIED LIFE SCIENCES, VIENNA (ABF-BOKU)
Peter Beigl
Stefan Salhofer

INSTITUTE OF ENVIRONMENT PROTECTION ENGINEERING
WROCLAW UNIVERSITY OF TECHNOLOGY (WUT)
Ryszard Szpadt
Marta Sebastian
Iwona Maćków
Paweł Mrowiński

LABORATORY OF PROJECT MANAGEMENT
DEPARTMENT OF CIVIL ENGINEERING
SCHOOL OF ENGINEERING
DEMOCRITUS UNIVERSITY OF THRACE (DUTH)
Konstantinia Tsilemou
Demetrios Panagiotakopoulos

NOVATEC S.À R.L. (NOVATEC)
Lothar Schanne

DEPARTMENT OF MUNCIPAL WASTE AND PUBLIC SPACE
SYNCERA DE STRAAT (DE STRAAT)
Clemens Berntsen

INFRASTRUKTUR & UMWELT (I&U), PROFESSOR BÖHM UND PARTNER
Orhan Boran
Umur Natus –Yildiz
Gernod Dilewski

SERVEI DE TECNOLOGIA QUIMICA, TECHNOLOGY TRANSFER SERVICE
UNIVERSITAT ROVIRA I VIRGILI (STQ-URV)
Francesc Giralt
Elena Suñé
Josepa Garreta
Eva Mª García

WASTE MANAGEMENT - ECOLOGY CONSULTING COMPANY
WAMECO S.C. (WAMECO)
Włodzimierz Szczepaniak
Wojciech Górnikowski

DEPARTMENT FOR ENVIRONMENTAL ENGINEERING
KAUNAS UNIVERSITY OF TECHNOLOGY (KTU)
Gintaras Denafas
Viktoras Racys
Ingrida Rimaityte

FACULTY OF CIVIL ENGINEERING ,DEPARTMENT OF SANITARY ENGINEERING
SLOVAK UNIVERSITY OF TECHNOLOGY (SUT-BA)
Katarina Jankovicova
Oskar Cermak

Contributions Listed by Project Partners

Chapter	Project Partners
1 Introduction	Emilia den Boer Jan den Boer
2 Waste Prognosis	Peter Beigl Jan den Boer
3 Assessment Tool	Emilia den Boer
4 Environmental Assessment	Emilia den Boer Jan den Boer
5 Economic Assessment	Konstantinia Tsilemou Demetrios Panagiotakopoulos
6 Social Assessment	Montse Meneses Julio Rodrigo Emilia den Boer Jan den Boer Jan Berger Orhan Boran
7 Temporary Storage	Peter Beigl Jan den Boer
8 Collection and Transport	Peter Beigl
9 Treatment	Emilia den Boer Jan den Boer
9.9 Waste Recycling	Montse Meneses Julio Rodrigo
9.10 Energy and auxiliary materials	Jan den Boer Julio Rodrigo
10 Case Studies	Orhan Boran Iwona Maćków
10.2.1 Xanthi-Greece	Konstantinia Tsilemou Demetrios Panagiotakopoulos
10.2.2 Kaunas-Lithuania	Gintaras Denafas Viktoras Racys Ingrida Rimaityte
10.2.3 Wrocław-Poland	Iwona Maćków Wojciech Górnikowski Ryszard Szpadt Marta Sebastian
10.2.4 Nitra-Slovakia	Katarina Jankovicova Oskar Cermak

Chapter	**Project Partners**
10.2.5 Reus-Spain	Francesc Giralt Josepa Garreta Sandra Colom Elena Suñé Eva Mª García
10.3 Summary of results of case studies	Ryszard Szpadt
11 Good practice in planning of MSWMS	Orhan Boran
12 Waste Prognostic Tool guide	Peter Beigl
13 MSWMS Assessment Tool guide	Montse Meneses Julio Rodrigo

List of abbreviations

biol. ODM	biologically degradable Organic Dry Matter
BOD	Biochemical Oxygen Demand
CEE	Central and Eastern European countries
DM	Dry Matter
EEA	European Environment Agency
EU	European Union
ESWM	Environmental Sustainability in Waste Management
EWC	European Waste Catalogue
GNP	Gross National Product
HDPE	High Density Polyethylene
IE	Inhabitant Equivalent
JRE	Java Runtime Environment
LBC	Liquid Beverage Cartons
LCA	Life Cycle Assessment
LCA-IWM	acronym of the Project: 'The Use of Life Cycle Assessment Tools for the Development of Integrated Waste Management Strategies for Cities and Regions with Rapid Growing Economies'
LCI	Life Cycle Inventory
LCIA	Life Cycle Impact Assessment
LDPE	Low Density Polyethylene
MBP	Mechanical-Biological Pre-treatment
MDR	Mixed Dry Recyclables
MRF	Material Recovery Facility
MSW	Municipal Solid Waste
MSWMS	Municipal Solid Waste Management System
NGO	Non Governmental Organisations
NMVOC	Non Methane Volatile Organic Compounds
ODM	Organic Dry Matter
OECD	Organisation for Economic Cooperation and Development
PAN	Peroxyacetylnitrate
PET	Polyethylene Terephthalate
PPP	Purchasing Power Parities
p.y.	per year
RDF	Refuse Derived Fuel
TS	Temporary Storage
USD	American Dollar

WEEE	Waste from Electrical and Electronic Equipment
WWTP	Waste Water Treatment Plant
VOC	Volatile Organic Compounds

1 Introduction

1.1 General

This Handbook compiles the results of the research project 'The Use of Life Cycle Assessment Tools for the Development of Integrated Waste Management Strategies for Cities and Regions with Rapid Growing Economies', in short: LCA-IWM. The project, which ran from September 2002 until August 2005, was financially supported by the European Commission. It is part of the Fifth Framework Programme (contract number EVK4-CT-2002-00087) in the following way:

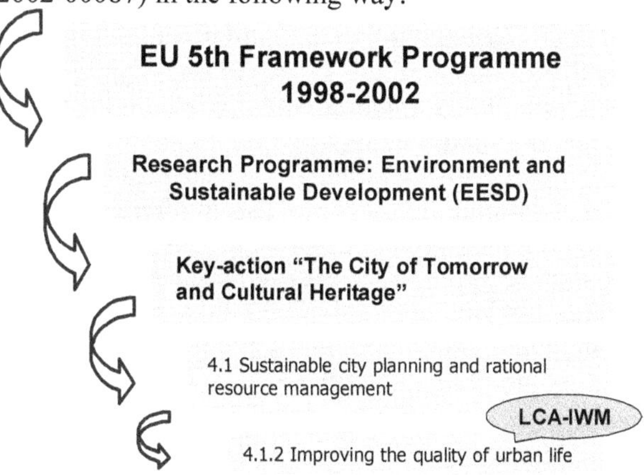

The contents of this Handbook however, do not necessarily represent the opinion of the European Commission.

The results of the LCA-IWM project consist of two decision support tools: a Waste Prognostic Tool and a Municipal Solid Waste Management System (MSWMS) Assessment Tool (also called: LCA-IWM Assessment Tool). The Prognostic Tool enables the prediction of future amounts of generated waste based on a limited amount of input parameters. These parameters consist of the current amount and composition of household waste and predictions of some general socio-economic indicators on the one hand and historical data on these factors on the other (available in the tool) . The Assessment Tool enables the planning and assessment of waste management strategies. Up to four different scenarios can be created and compared. Each scenario, apart from general user inputs, consists of three basic waste management sub-systems:

- Temporary Storage
- Collection and Transport
- Treatment, Disposal & Recycling

For all scenarios the environmental, economic and social impacts can be determined, providing a sustainability assessment of the various alternatives.

An introduction is given in Chapter 1. Chapters 2 and 3 provide the theoretical backgrounds of the Prognostic Tool and the Assessment Tool. In Chapters 4, 5 and 6 the Assessment Tool's approach is given for the environmental, economic and social assessment of waste management systems respectively. In the following chapters the calculation methods for the Assessment Tool's modules are provided: Chapter 7 for Temporary Storage, Chapter 8 for Collection and Transport and Chapter 9 for Treatment, Disposal and Recycling processes. Chapter 10 shows the results of the first use of the tools in 5 case-studies in European municipalities. Chapter 11 provides good practices in waste management planning. Finally Chapter 12 consists of user-guides for the developed Waste Prognostic Tool and the MSWMS Assessment Tool.

At the project's web-site both tools can be downloaded:
www.lca-iwm.net

Apart from the tools some of the project deliverables are also publically available. These are Deliverable 2.1 (waste prognosis); Deliverable 2.2 (waste prevention); Deliverable 3 (environmental assessment), Deliverable 4 (economic assessment) and Deliverable 5 (social assessment). The deliverables provide an in depth description of the theory and backgrounds of the Prognostic and Assessment Tools as well as detailed calculation methods. They are to be considered as annexes to this book, but are not offered in a printed version.

1.2 Municipal waste definition

In the European Waste Catalogue (EWC) a specification of municipal waste is given (EWC 2000/532/EC). The EWC category no. 20 "Municipal wastes and similar commercial, industrial and institutional waste including separately collected fractions" consists of:

20 01: Separately collected fractions: paper and cardboard, glass, small plastics, other plastics, small metals, other metals, wood, organic kitchen

waste, clothes, textiles, different types of hazardous waste (solvents, acids, pesticides, etc.);
20 02: Garden and park wastes: compostable waste, soil and stones and other non-compostable wastes;
20 03 Other municipal wastes: mixed municipal waste, waste from markets, street cleaning residues and septic tank sludge.
Additionally, in category no. 15 "Waste packaging; absorbents, wiping cloths, filter materials and protective, clothing not otherwise specified" point 15 01 addresses packaging waste, including separately collected municipal packaging waste. These include the following packaging waste: paper and cardboard, plastics, wood, metal, composites, mixed packaging, glass, textile and packaging contaminated by dangerous substances.

Within the LCA-IWM project primary focus is placed at management of **household and similar to household waste**. The definition of household waste varies from country to country. The European Environment Agency (EEA) defines household waste as solid waste composed of garbage and rubbish, which normally originates from houses (EEA 2003a). The wastes similar to household waste are those which originate from other sources than household, like offices and shops, and are collected and disposed of in the same manner as household waste.

1.3 Waste composition

In the LCA-IWM project the Waste Prognostic Tool is used to predict future quantities and composition of waste by main fractions. Table 1 summarises the types of waste which have been considered in the Prognostic Tool.

It should be born in mind that the Prognostic Tool provides data on waste generation, i.e. potential maximum quantities for separate collection. Within the Assessment Tool in the "Temporary Storage" module the actually collected fractions are specified by a user. The latter might differ from waste fractions described above, e.g. the user may design separate collection of packaging waste or mixed dry recyclables which can include paper, glass, plastics and composites and metals or any combination of these.

Table 1 Considered types of generated municipal solid waste

Material-related groups	Considered waste types
Recyclables	Paper and cardboard
	Glass
	Metals
	Plastics and composites
Organic waste	Bio-waste
	Garden waste
Hazardous waste	Hazardous waste
	Waste electrical and electronic equipment (WEEE)
Other materials	Mixed[a] or Residual[b] waste
	Bulky waste

[a] Mixed waste is understood as household and similar to household waste collected in a municipality where no separate collection scheme is implemented.
[b] Residual waste consists of those wastes which remain for collection after source separation of recyclables.

Modelling of treatment processes requires further specification of waste fraction into individual materials. Material composition of separately collected fractions is provided in Section 9.9. Data for material composition of fractions that are collected within the residual/mixed waste stream are given in Table 2. It should be noted that the material composition in separately and non-separately collected fractions varies. It happens because some materials are easier sorted than the others.

Table 2. Default values for material composition of waste fractions within household mixed/residual waste (Dehoust et al. 2002)

Waste fraction	Waste materials	Material contribution [%]
Metals	Iron	91
	Aluminium	9
Plastics and composites	Plastics	56,1
	Packaging composites	5,4
	Composites	38,5
Residual/Mixed other waste (different than already listed)[a]	Diapers	20,8
	Inerts	13,2
	Textiles	7,8
	Leather	1,3
	Medium corn[b]	28,3
	Fine corn[c]	28,6

Waste fraction	Waste materials	Material contribution [%]
WEEE	Iron	52
	Aluminium	7
	Copper	9
	Plastics	22
	Glass	7
	Inerts	3

[a] Residual/Mixed waste consists partly of the separately collected fractions (organic waste, paper, glass, etc.) which were not separated from the waste and partly of other materials which are not included in any separate collection scheme. The latter are called here: "Residual/Mixed other waste" and include the following materials: diapers, inerts, textiles, leather, medium corn (8 – 40 mm) and fine corn (< 8 mm).
[b] Medium corn was assumed to consist in 49% of organic matter and in 51% of inerts (Rotter 2004).
[c] Fine corn was assumed to consist in 39% of organic matter and in 61% of inerts (Rotter 2004).

Composition of bulky waste is provided in the Table 3. Due to lack of data no distinction is made here between separately collected bulky waste and bulky waste within mixed/residual fraction.

Table 3. Default values for material composition of bulky waste (Dehoust et al. 2002)

Waste fraction	Waste materials	Material contribution [%]
Bulky waste	Wood	49
	Iron	8
	Paper	2
	Inerts	4
	Plastics	12
	Medium corn	20
	Textiles	5

2 Waste Prognosis

2.1 Introduction

Appropriate decision making in waste management requires knowledge about the impacts on present and future waste generation and separate collection. Previous estimations without extensive consideration of waste generation characteristics have resulted in an expensive overcapacity for waste treatment facilities. Particularly in countries lacking high environmental standards, but with rapid economic development, these impacts are both remarkable for and relevant to waste generation prognosis as well as the assessment of the future separate collection performance. Therefore the problem to be solved requires an accurate method of identifying relevant factors of influence in terms of waste quantity development.

In order to provide quantitative estimations in appropriate accuracy, two consecutive, but independent estimations had to be carried out. These cover the estimation of

- future waste generation (regardless of whether a fraction is collected separately or not) and of
- the separate collection performance in future (which percentage of a waste fraction (e.g. glass) is collected separately?).

Figure 1 shows the relationship between the waste generation of two fractions in a simplified way (left column), e.g. paper, which can be partly collected separately at the source (right column) and collected unseparated as a fraction within the residual waste (middle column).

The future waste generation is based on analyses of the relationship between socio-economic conditions and the waste generation rate. Their estimation is explained in Section 2.2.

The rate of separate collection (e.g. percentage of source separated paper as percentage of total generated paper) depends on the citizens' behaviour and the feasibility for the municipality to promote separated collection. The analysis is shown in Section 2.3.

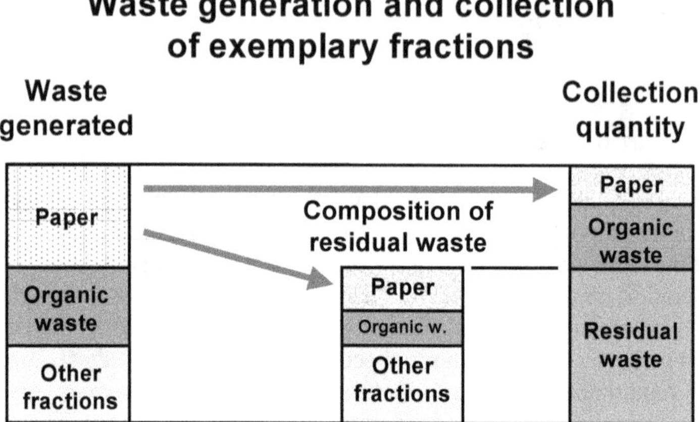

Figure 1. Waste generation and collection of exemplary fractions.

2.2 Prognosis of municipal solid waste generation

The generation of municipal solid waste is the starting point within each waste management system. The input covers a wide set of products which are bought, used, transformed, stored and later disposed of by the waste generator (Figure 2). The output can be attributed to groups with similar material characteristics like paper and cardboard, organic waste etc.

The amount of solid waste generation per capita strongly depends on the social and economic conditions. This can be proved by remarkable differences in the MSW generation rates in European cities. As an example, a comparison of economic areas in the year 2000 shows that major European Union-15 (EU-15) cities were characterised by far higher Municipal Solid Waste (MSW) generation rates (510 kg/cap/yr) than the Central and Eastern European countries (CEE) cities (354 kg/cap/yr).

The future growth of waste quantities is of even more higher interest (than the amounts themselves) as its estimate is an important prerequisite for adequate capacity planning of waste management systems. Inappropriate estimates can lead to higher costs and environmental burdens due to e.g. overcapacity of processing facilities. The accurate estimation can be even complicated in case of areas with similar economic status (though it could be expected that this comparison would be easier than in case of heterogeneous areas). Major CEE cities can serve as example: Here the annual growth rate of per-capita MSW generation between 1995 and 2001 ranges between a reduction –5% p.y. (per year) and up to more than +10% p.y., while the average value lies at a growth rate of +4.3% p.y.

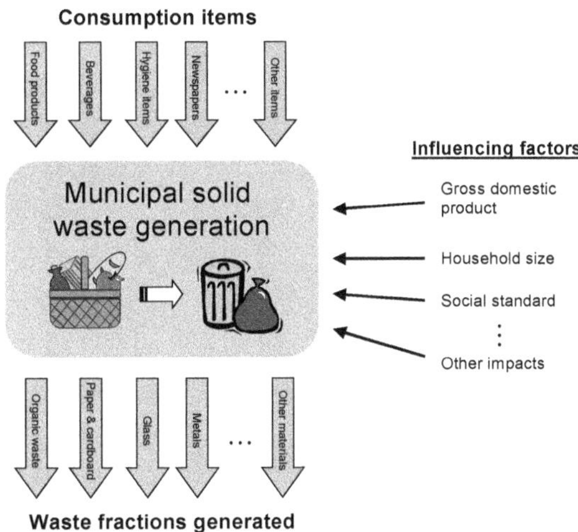

Figure 2. Municipal solid waste generation.

In the following the background and procedure of the applied waste prognosis method will be described. All calculation details can be found in Deliverable 2.1.

What is the benefit of advanced waste generation prognoses?

If not even made by rule of thumb, the usually practised way of MSW forecasts considers the population growth as well as the long-term development of per-capita generation which will be simply extrapolated over time. Other significant indicators on the underlying social and economic developments are normally neglected by the mainly technically skilled experts who are concerned with this topic (Karavezyris 2001). Figure 3 shows a rather extreme example for a misleading trend extrapolation leading to an error of approx. 100 kg/cap/yr within only three years.

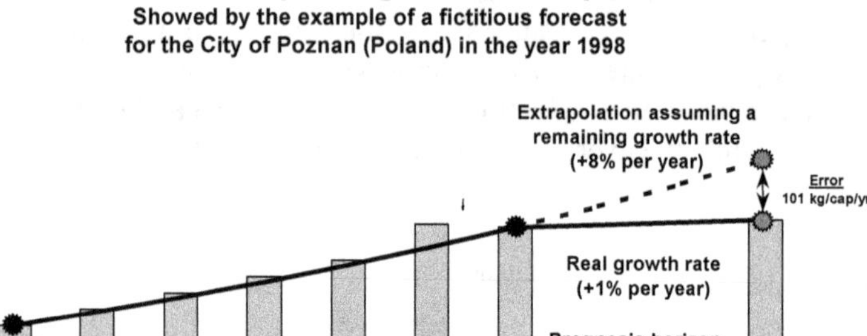

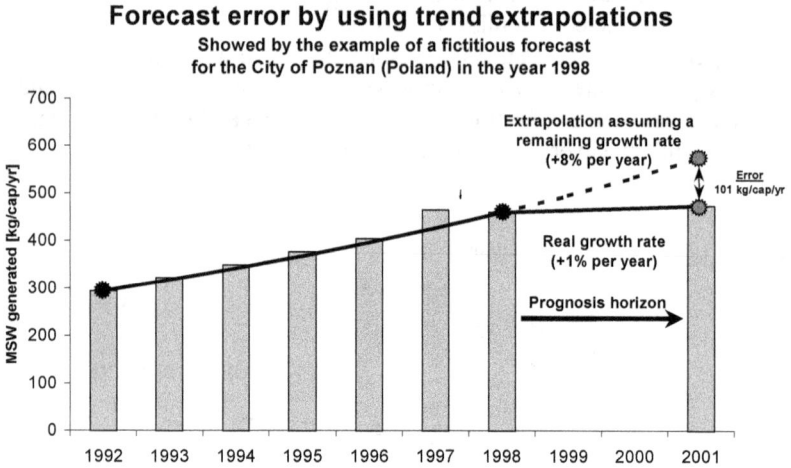

Figure 3. Forecast error by using trend extrapolations – Example.

Although advanced prognosis modelling cannot totally prevent forecast errors (which normally increase with rising prognosis horizons), the consideration of a set of socio-economic indicators will lead to substantial reductions of occurring forecast errors (see Armstrong et al. 2001). The application of these indicators within such models can be objectively tested with historic time series of both waste generation data and the parameter data (details about the tests and the performance indicators of the here applied model in the LCA-IWM tool is described in Deliverable 2.1.).

Europe-wide investigation of potential factors and waste-related data

In order to find out relationships between potential influencing factors and municipal solid waste generation, a Europe-wide investigation was carried out within this project.

The investigation covered the collection and inspection of

- waste-related data as well as
- data in terms of economic, demographic and social indicators (their selection based on previous studies concerning this topic in all major European cities with more than 500,000 inhabitants.)

The project partners co-operated with local city representatives who provided waste-related and socio-economic data at the city level. Additionally, national data were obtained from international organisations, such as the United Nations or Organisation for Economic Cooperation and Development (OECD). To enable the analysis of developments in time, the collection of data covered the years from 1970 to 2001.

Finally, it was possible to collect data about the total municipal solid waste quantities (including general data at the city and country level) in 55 major cities in the EU-15 and 10 CEE countries (out of a total of 65 respectively 91 cities in the whole of Europe including Turkey) with an average time-series length of ten years. In terms of the waste generation of the main materials (paper and cardboard, glass etc.) only 45 data sets from 31 cities were available.

Influencing factors on waste generation

Based on the created database, it was possible to confirm or reject the influence of the investigated indicators on the quantity of waste generated by means of a couple of statistical methods (details see Deliverable 2.1). The goal was the identification of significant factors on

- the total municipal solid waste generated per capita as well as
- the composition of MSW by main fractions (in mass-%), namely paper and cardboard, glass, organic waste, metals and plastics and composites.

More than 35 potential influencing factors have been selected for the mentioned investigation. Due to missing data and partly implausible data quality (caused e.g. by non standardised definitions), 14 indicators were available on city and national level (Table 4) which were evaluated in the following.

Table 4. Available indicators at city and national level

Available indicators at city and national level	
• Total population	• Population density
• Population age structure (0 to 14 years / 15 to 59 years / 60 and more years)	• Sectoral employment (agriculture / industry / services)[a]
• Gross domestic product	• Infant mortality rate
• Overnight stays	• Life expectancy at birth
• Average household size	• Unemployment rate

[a] Only available (and also useful) on national level

The analyses resulted in the identification of a set of significant indicators. The selection of these based (beside the primarily necessary statistical significance) on the availability as well as predictability which should ease the usability for the user. In the following several factors are described which have a significant influence on the MSW generation and are thus used as model parameters (see Table 5):

- Gross domestic product: This indicator of the economic power of a region has often been employed to assess the waste stream development. Such a relationship was confirmed especially for very rich cities (See Deliverable 2.1).
- Social indicators: Social affluence indicators had not been considered in waste stream assessments in the past. In this model, however, infant mortality rates, life expectancy and the percentage of persons employed in agriculture were shown to have a significant impact, especially in rapidly growing countries in Eastern Europe. These indicators are advantageous because of the accessibility and quality of the data and predictability.
- Age: The positive relationship between the group of people aged 15 to 59 years (i.e. the age group with highest economic activity) and the amount of generated waste confirms previous experience (Sircar et al. 2003).
- Household size: A small number of people living in an average household is also an indication that more waste is produced (also refer to Dennison et al. 1996).

Table 5. Influencing factors as model parameters on MSW generation.

Factor	Unit	Impact
Gross domestic product per capita[a]	USD PPP[b] at 1995 prices	+
Infant mortality rate	Per 1.000 births	–
Population aged 15 to 59 years	Percentage of total population	+
Household size	Persons per household	–
Life expectancy at birth	Years	+
Labour force in agriculture	Percentage of total labour force	–

[a] Only the national indicator is significant in these cases.
[b] American Dollar Purchasing Power Parities

The mentioned relationships have been condensed in a model which contains a set of formulas. The model enables the prediction of the per-capita MSW generation with an accuracy of 8% ($R^2 = 0.65$) - this error is significantly lower than that achieved in similar models (compared with 14% prediction error for German cities (Sircar et al. 2003) or an $R^2 = 0.49$

for the prediction of waste paper in Austria (Bach et al. 2004)). The description of the calculation procedures (e.g. model coefficients, applied methods, testing procedures and validation) is shown in Deliverable 2.1.

Estimating future waste generation

The estimation of future waste generation is similar to the above mentioned model which is able to quantify the relationship between socio-economic conditions in an area and the amount of waste generated.
The applied prognosis model is based on the following assumptions:

- Similarity of relationships between factors and waste generation: It has been assumed that the relationships between socio-economic parameters and waste generation are similar for all cities
  - in the analysed past years (from 1970 up to 2001 -- based on the mentioned evaluations) and
  - in the future years (which have to be projected).
- Primary use of projections of model parameters on city level: the most accurate projections of social or economic indicators (which are necessary for future waste estimations) can be provided by the concerned region itself. This means that the most appropriate projections, e.g. for the population growth or household size in a city, can be best provided by the municipal offices (and not by national or supranational organisations) which have the better knowledge of the regional peculiarities. Therefore it has been assumed that the development of the necessary indicators in time – actual and projected future values – should be primarily provided by the municipality itself. If the user is unfortunately unable to provide an indicator value for a city, the model suggests an appropriate national value as proxy for the city indicator (details see next section).
- Additional use of national projections: Along with city-related parameters, also national parameters are necessary for the waste prognoses (see Table 6). Actual values and projections of international organisations will be recommended by the tool as default values which are based on an extensive background database within the model (see next section).

The simplified procedure of waste generation estimation is as following:

1. The user inputs actual waste generation and composition data (only for waste composition data default values are available for the user).
2. Socio-economic indicators of the actual year have to be provided by the user or, if not available, as model default values.

3. Socio-economic trends up to the assessment year are input by the user or, alternatively provided as model defaults.
4. The model then calculates the estimated MSW generation (by fractions) on the basis of the projected model parameters.

The model has been tested with historic time series of the collected database. The resulting forecast error (median values) increases with enlarging prognosis horizons and amounts to

- 5.3% for time series with 5 to 10 years and
- 7.7% for time series with 11 up to 22 years.

These moderate errors show the benefit of this model as a helpful decision support tool in comparison with usual methods.

Details concerning the calculation procedure are available in Deliverable 2.1.

Provision of socio-economic background data on national level

It is not always possible for the user to obtain the necessary input data. For this case an extensive database of socio-economic data from 32 European countries has been implemented in the tool. Table 6 shows the list with the database for indicators, both with historic data as well as projections which have been carried out by renowned international institutions. Table 7 shows the list with countries for which this data is existing.

Table 6. Background database for European countries.

Indicator	Covered period	Sources
Total population	1990 – 2030	UNECE, 2004
Infant mortality rate	1990 – 2025	UN-ESA, 2003
Population aged 15 to 59 years	1990 – 2030	UNECE, 2004
Household size	1990 – 2025	UN-Habitat, 2004
Life expectancy at birth	1990 – 2025	UN-ESA, 2003
Economically active persons in agriculture	1980 – 2010	FAO, 2004
Gross domestic product – historic	1990 – 2001	OECD, 2002
Gross domestic product – projections	2002 – 2015	European Union, 2004; World Bank, 2003

Table 7. Countries with existing background data.

Countries with existing model background data (according to Table 6)			
Austria	France	Lithuania	Romania
Belgium	Germany	Luxemburg	Slovak Republic
Bulgaria	Greece	Macedonia	Slovenia
Croatia	Hungary	Malta	Spain
Czech Republic	Ireland	Netherlands	Sweden
Denmark	Iceland	Norway	Switzerland
Estonia	Italy	Poland	Turkey
Finland	Latvia	Portugal	United Kingdom

Municipal waste prevention measures and their expected impact

In the framework of this project, the term 'waste prevention' is defined as quantitative reduction of overall municipal solid waste quantities. Prevented waste quantities do not affect the municipal waste collection and treatment and thus do not cause environmental burdens in each stage of the waste management system.

Matters of interest within the framework of this project are waste prevention measures on municipal level (i.e. national measures are neglected, because not influenceable by cities). In general the influence of municipalities on the behaviour of citizens is limited. An extensive study of possible measures of waste prevention, carried out in Vienna (Salhofer et al. 2000), led to the result that out of 22 investigated prevention methods, only six possessed a sufficient potential and were feasible from the legal viewpoint by the city administration. It was estimated that altogether, about 30 kg/cap/yr waste could be avoided by the application of these 6 measures, of which approximately 15 kg/cap/yr were residual waste (bagged waste), corresponding to approximately 5% of the total Viennese residual waste quantity.

Experiences of this and similar studies about this topic will provide benchmarks for the quantitative prevention effect of these measures. In cities with developed economy, high incomes and appropriate consumption behaviour of the inhabitants, waste prevention potentials can be initiated by municipal programs, if the required measures are pursued rigorously. The approximate quantitative effect of a set of prevention measures is shown in Table 8. The prevention effects are represented as mass percentage of the considered waste quantities (related to potentials or collection fractions). More details about these waste prevention measures are available in the Deliverable 2.2 'Policy recommendations for waste reduction measures and their expected quantified impacts'.

The following design options are provided for the implementation of municipal waste prevention measures. The user has the following choices:

- Are there plans to implement prevention measures in general? (yes/no)
- Implementation period for these measures: The planned start and final year of implementation has to be input. It is recommended to implement the measures between a span of 2 to 6 years.
- Implementation rate for each measure: The implementation rate is defined as the percentage of citizens (for public measures) respectively percentage of municipal officers (in case of the internal measures) which can be targeted by each measure (0%, if a measure will not be implemented). It is notable that an implementation rate of 100% is very difficult to achieve due to cost constraints. The implementation rate concerning the intensification of public relations is feasible only in small districts of a city (e.g. single block of flats) and thus limited to maximum 20%.

The quantitative effect of these measures will be considered in the following way. The prevention potential of each measure (the assumed mass percentage of the paper and cardboard potential, organic potential or total MSW generated (see Table 8) multiplied by the implementation rate by each measure) will be deducted from the waste quantity in a defined year which has been estimated by means of the waste prognosis approach mentioned above. During the years of this implementation period, it has been assumed that prevention effects are linearly growing from the start year till the final year of implementation.

Table 8. Approximate waste prevention potentials by measure

Waste prevention measures at municipal level	Prevention potentials by fraction [mass-%]		
	Paper and cardboard potential[a]	Organic waste potential[a]	Total municipal solid waste
Public measures			
Mailbox sticker "no junk mail"	9.52		
Promotion of repair services (Guide)			1.32
Promotion of reusable goods (Guide)			0.11
Promotion of hire services (Guide)			0.04
Promotion of nappy service (Subsidies)			0.28
Intensified municipal public relations			5.00
Promotion of the do-it-yourself composting		11.17	
Internal measures			
Procurement of reusable material			0.09
Double-sided use of paper	0.63		
Application of reusable towels			0.04

[a] Potential = Source separated quantity + Quantity of fraction within residual/mixed waste

2.3 Separate collection performance

Separate collection by waste generators (see Figure 4) plays a central role in integrated waste management. The way of collection of waste types determines a set of possible waste management options, like materials recycling, biological treatment etc. (McDougall et al. 2001).

In the Prognostic Tool the amounts of generated waste are determined for all concerned fractions in the year of assessment. There are a variety of factors which influence which share of the generated waste is collected separately. Since the separation of the waste is solely undertaken by the inhabitants of a considered city, the citizens' behaviour is the key influencing factor.

Figure 4. Waste collection at source.

Poiesz (1999) states in the so-called Triade model, that citizen behaviour is influenced by three factors. Zegwaard (2000) applied this theory to waste separation:

- motivation (are citizens motivated to separate waste?)
- capacity (are citizens physically and mentally capable to separate waste?)
- opportunity (are opportunities provided to enable citizens to separate waste?)

In Figure 5 the influence of the three citizens' behaviour factors is shown. If any of these factors is minimal, the resulting waste separation behaviour will also be minimal. Traditionally, municipalities concentrate on optimising the opportunity of the citizens, by offering the technical equipment of waste management systems. Bins and containers are offered and collection services provided for various fractions. The motivation and capacity of citizens are often less focused upon.

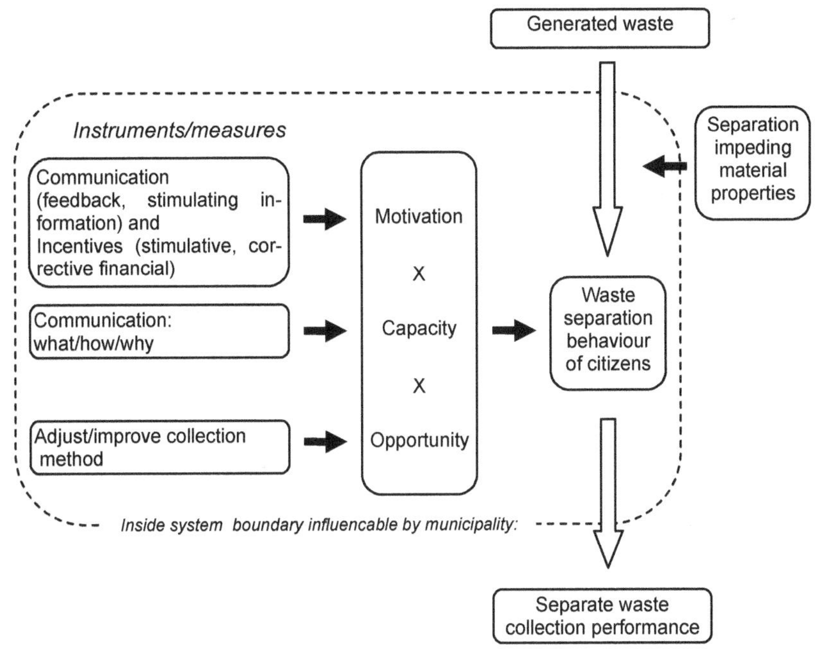

Figure 5. The Triade model applied to waste separation behaviour

In the LCA-IWM project the separate waste collection performance in European cities has been analysed. Since a quantification of the above mentioned factors to make a prediction of the waste collection performance is not realistic, a different approach was followed.

Europe-wide investigation of source-separated collection rates

The realistic estimation of achievable rates of source-separated collection quantities of recyclables can be better ascertained on the basis of documented experiences rather than through theoretical approaches. Thus a Europe-wide investigation concerning the collection performance of different waste management systems was carried out.

It has been assumed that experiences about the development of collection rates in mature collection systems (e.g. in Germany or Austria) can be transferred to developing collection systems with actually low rates of separately collected fractions.

Based on the investigation mentioned in Deliverable 2.2 the following data amongst other aspects has been collected:

- collection quantities of separated and non-separated waste fractions
- documented sorting analyses of residual waste
- general socio-economic status of the city

The collection performance was deduced from this data by dividing the separated collection quantities by the total generated quantity of this fraction (i.e. separated fraction and fraction within the residual waste (data from sorting analyses)).

It has been proved (details see Deliverable 2.2) that the source-separated collection rates in the cities clearly depend on the general socio-economic status. Four prosperity-related levels (classification see Deliverable 2.2) have been defined whereby the socio-economic development of certain cities in time could be facilitated to compare with realistic, feasible benchmarks for collection quantities (details to the grouping see Deliverable 2.2).

Achievable target and optimum values for separate collection

Based on the mentioned prosperity-related grouping of cities, it has been assumed that the average value of collection rates of all cities in the wealthiest prosperity group (this group contains cities in Germany, France, Ireland, United Kingdom, Italy and Netherlands) should serve as 'target value' for the remaining cities which should be achieved within 10 years after implementation of a separate collection scheme. Alternatively the achievement of the highest percentile in this group should serve as 'optimum value'.

Table 9 shows the calculated target and optimum values expressed as mass percentage or per-capita generation rate for four waste fractions.

Table 9. Recommended target and optimum values for separated collection.

Waste fraction	Recommended rates of separate collection				Recommendation of Öko-Institut e.V. (1999)
	Target value		Optimum value		Target value
	[mass %]	[kg/cap /yr]	[mass %]	[kg/cap /yr]	[kg/cap/yr]
Paper and card-board	45	50	74	83	90
Glass	50	22	69	30	40
Plastics and com-posites	33	19	65	39	30
Bio-bin collected organics	22	35	51	82	100

Implementation of source-separation programs in municipalities

The user of the tool has the following design options concerning the planned implementation of separate collection in each sector of the city:

- no implementation is planned (and none is currently existing)
- separated collection is already in existence to the required extent
- implementation is planned in the future: In this case a future year (between the following year and the assessment year) can be selected

The calculation of the impact of the collection performance (and also eventually waste generation which is deviating from the generation rate of the whole city) is presented in Deliverable 2.2.

3 Assessment Tool

The LCA-IWM Assessment Tool is a decision support tool for waste management planning. The tool allows modelling of waste management scenarios at a municipality level. Targeted end-user is a municipal officer responsible for waste management planning, who appreciates having more insight into potential impacts of his decisions. For this purpose modules representing individual waste management processes, such as Temporary Storage, Collection, Transport and Treatment have been developed. Using these modules the user designs scenarios which he wants to consider for his municipality. The assessment part of the tool consists of environmental, economic and social assessment systems. The bases for assessment are sustainability criteria and their quantitative indicators, which are computed in the tool. Thus the assessment part contains algorithms allowing calculation of environmental, economic and social performance of a given scenario. Necessary inputs for calculation of performance indicators are inventory data such as e.g. emissions for the environmental assessment and costs for the economic assessment. These inputs are derived from waste management processes modules. The indicator results are aggregated for each scenario and based on the final result the user can select optimal scenario for his particular municipality.

In the Assessment Tool a large number of input data is needed. For most of these the tool already contains values. Hereby there are two levels of default values. For most input values a default is offered in the Assessment Tool's interface. It can be used in case the tool-user cannot provide its own value. In the corresponding modules which are running in the tool background the Assessment Tool there are many default values, which can only be changed by experienced users.

3.1 Scope of the assessment and functional unit

The developed methodology within the LCA-IWM project provides means for assessment of alternative municipal solid waste management systems (MSWMS). The borders of assessment are extended to include the environmental, social and economic impacts occurring at all stages of a waste management system, i.e.: Temporary Storage of waste, Collection, Transport, Waste Treatment and Final Disposal. Figure 6 presents the boundaries of environmental assessment. It shows the stages of waste management and individual processes which are considered in this study.

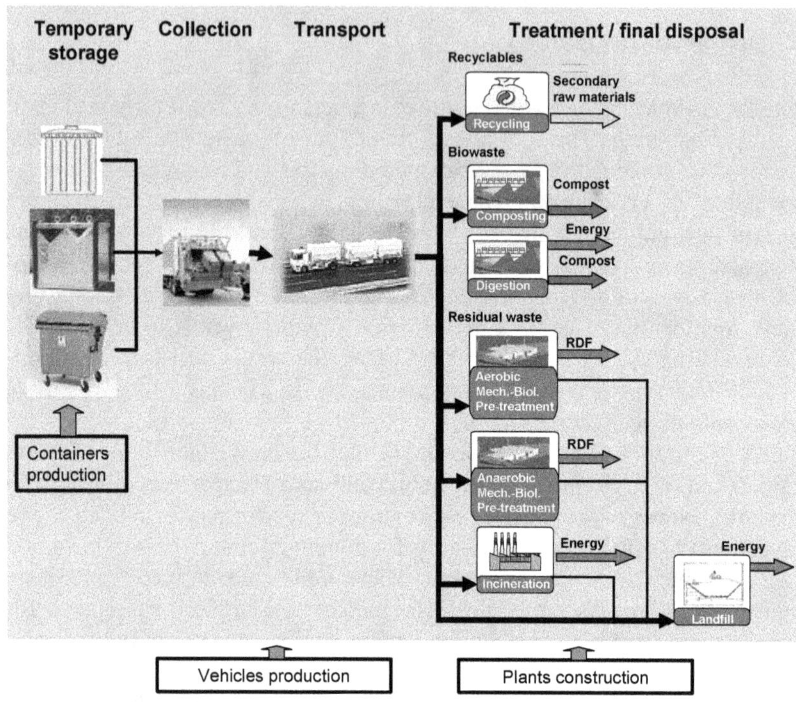

Figure 6. The boundaries of environmental assessment

The assessment starts at the moment waste is put in a Temporary Storage system (bags, bins and containers). From there the waste is collected and transported to the Treatment/Disposal site. Potential Treatment options are Recycling of separately collected recyclables, Composting of biowaste, aerobic and anaerobic Mechanical Biological Pre-treatment (MBP) or Incineration of residual waste. Apart from waste flows, also the flows of products resulting from waste processing, such as secondary materials obtained from recyclables, compost derived from organic waste and energy produced in the waste treatment processes are considered in the assessment. The production of those products is accounted for as a positive effect, so called "credit" (see Figure 7). The environmental impact assessment is concerned with emissions of pollutants and resources consumption throughout the system. In the economic assessment a substantial cost factor is the investment cost. This includes investment in waste containers, collection and transport vehicles and treatment plants. In the environmental and social part construction of plants and production of waste management equipment is considered of far less importance than the operation of the system. Thus the impacts arising at the construction phase are excluded from the assessment. The only investment included in the environmental assessment is the impact of production of bags, bins and waste containers. This is due to a short life time of these items.

The functional unit of the proposed assessment method is the amount of waste generated in a city and entering the waste management system within one year.

3.2 Data inventory and allocation procedures

The intention of the study is to arrive at an assessment method for waste management in rapid growing European cities and regions, i.e. areas where waste management is not yet very advanced. Therefore, generic data sets on process efficiencies, costs, resources consumption and emissions from waste treatment/disposal processes are used. The data stem from countries with fairly advanced waste management technologies.

Allocation of flows and releases is based on allocation of credits for the additional co-products. The crediting system allows to account for benefits of valorisation of waste. The additional co-products, such as electricity, heat, refuse derived fuel (RDF) and recycled secondary materials substitute the products generated in conventional technologies. Thus credits in form of negative energy and material flows equivalent to the quantities of replaced primary products are assigned to respective processes. For example in the process of waste incineration apart from disposal of waste a co-generation of electricity and heat takes place. Thus for every kWh of electricity and kJ of heat a credit in the form of negative flows related to production of one kWh of electricity and one kJ in a power plant are assigned.

The co-products of waste considered in the environmental assessment management modules are illustrated in Figure 7.

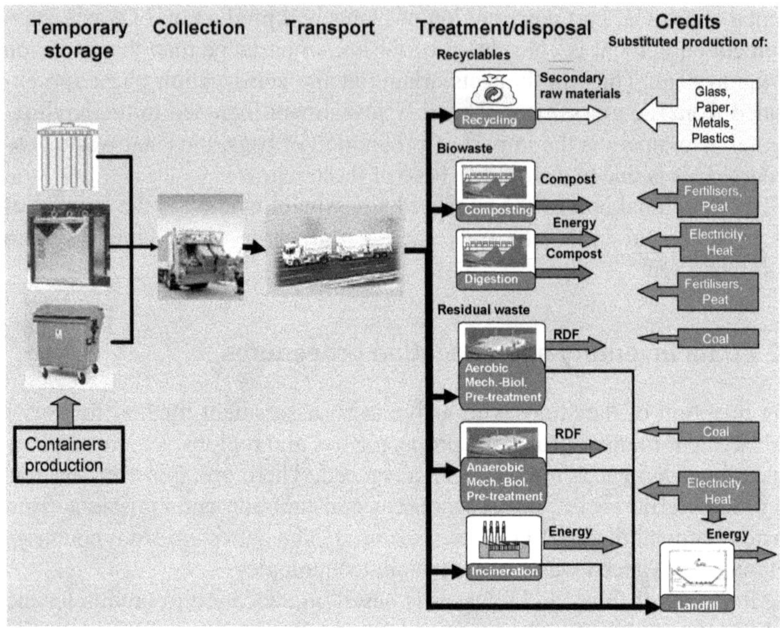

Figure 7. Credits for co-products of waste management system

4 Environmental Assessment

In "Our Common Future", the Brundtland commission proposed the now often-cited definition of sustainable development (WCED 1987):
"Sustainable development is development that meets the needs of the present without compromising the ability of future generations to meet their own needs".

The commission suggested that a strategic framework to integrate economic and environmental factors in policy decisions was needed at national and international levels. Based on this, efforts have been undertaken to develop a single definition of sustainability. Since it has been difficult to adopt a single definition of sustainability in an environmental, economic and social understanding, individual definitions for these three aspects of sustainability are often proposed. Also within this project it has been decided to consider the three sustainability aspects on an individual basis. Thus in this chapter only environmental aspects are referred to.

4.1 Introduction

One of the broadly recognised definitions of environmental sustainability is the one provided by Goodland after Daly (Goodland 2002):
"Environmental sustainability itself seeks to improve human welfare by protecting the sources of raw materials used for human needs, and ensuring that the sinks for human wastes are not exceeded, in order to prevent harm to humans. (...) On the sink side, this translates into keeping emissions within the assimilative capacity of the environment without impairing it. On the source side, harvest rates of renewables must be kept within regeneration rates ..."
This definition serves as a basis for sustainability assessment in waste management within this study.

4.1.1 Environmental sustainability in waste management

Based on the above provided definition of environmental sustainability, general objectives for any human activity can be summarised as an objective of rational resource consumption and reduction of environmental pollution. Hence, also Environmental Sustainability in Waste Management (ESWM) may be expressed through these two major objectives

- conservation of resources and
- pollution prevention.

Within this report ESWM will be further specified through defining assessment criteria and quantitative indicators to measure environmental sustainability of alternative waste management scenarios.

4.1.2 Targets of Waste Management according to the European waste policy

Notwithstanding the environmental sustainability criteria, municipal Waste Management Systems must comply with the general requirements of European waste policy. For waste management planning at the municipal level the most relevant directives are Framework Directives on Waste and on Hazardous Waste (European Council Directives: 75/442/EEC and 91/689/EEC,) and Community Strategy for Waste Management (European Council Resolution: COM (96) 399 final) as well as directives on Packaging and Packaging Waste (European Parliament and Council: 94/62/EC and 2004/12/EC), Landfill of Waste (European Council: 99/31/EC) and Waste Electrical and Electronic Equipment (European Parliament: 2002/96/EC). For planning of new waste management facilities Incineration and Landfill directives (European Parliament: 2000/76/EC and 99/31/EC, respectively) apply.

The main objectives of The Waste Framework Directive (75/442/EEC) and EU Strategy for Waste Management with respect to municipal waste management planning can be summarised as follows (ETC-WMF 2003):

- to secure the conservation of nature and resources, waste generation must be minimised and avoided wherever possible (prevention principle)
- to secure a reduction of the impacts from waste on human health and the environment, especially to reduce the hazardous substances in waste through the precautionary principle;
- to establish an integrated and adequate infrastructure by establishing an integrated and adequate network of disposal facilities based on the principle of proximity and self-sufficiency;
- to apply a hierarchy of waste management operations, which gives waste prevention the highest priority, followed by recycling and other types of recovery and landfilling as the least favourable option.

The Directive on Hazardous Waste stipulates development of waste management plans for the management of hazardous waste.

The Directives on Packaging and Packaging Waste, Landfilling and WEEE prescribe specific (quantitative) targets which have to be fulfilled by Member States:

- packaging recovery and recycling targets,
- targets for diversion of organic waste fraction from landfilling,
- collection and recovery targets of waste of electric and electronic equipment (WEEE).

4.1.3 Sources of criteria for environmental sustainability assessment of municipal waste management

The assessment criteria proposed within the LCA-IWM project are based on the defined environmental sustainability and European waste policy targets. The main environmental sustainability targets for waste management as derived from the environmental suitability definition are in compliance with the targets of the European Framework Directive on Waste. Compliance with the principles and targets of the European waste policy as well as national regulations is obligatory for any planned waste management system. The problem remains how to attain regulatory compliance in an optimal way. In an early planning phase not much data about system performance is available and the quantification of its impacts is hardly possible. Thus modelling and analysis of scenarios enables prediction of the system performance against the regulatory targets. The European Framework Directive on Waste prescribes resources conservation and reduction of impact from waste management to the environment and human health. In the past years a number of environmental assessment methods have been developed to describe and quantify these environmental interventions. One of such methods is Life Cycle Assessment (LCA). LCA provides a quantified assessment of environmental aspects in terms of pollution (emissions) and resources consumption over an entire life cycle of a given product/service. Life Cycle Impact Assessment (LCIA) methodology allows the quantification of potential impacts of these environmental interventions on human health and the environment. The principles of LCA and LCIA are explained in more detail in Chapter 3 of the Deliverable 3. As opposed to the other assessment methods, such as Environmental Impact Assessment or Risk Assessment, LCA allows comparisons of different systems based on a generic, i.e. not site specific, data. LCA is capable of comparing environmental burdens caused by different systems, but makes no reference to what will be the actual impact in a specific site. This means that the site sensitivity and the state of the environment in the given location are not considered. Thus, caution should be taken while analysing the outcomes of LCIA and the calculated environmental impacts should be considered as potential impacts. On the other hand, LCA is suitable for comparing alternative waste management scenarios for a given location according to the defined ESWM and general EU waste policy objectives (protection of the environment and human health and resources consumption). Therefore, LCA methodology has been selected as basis for development of environmental sustainability criteria for the purpose of this project.

LCA methodology does not allow for direct assessment of the adherence to the other principles of EU waste management, such as waste prevention and hierarchy of waste treatment options. LCA methodology also does not have capacity to deal with implementation of specific recovery/recycling targets. Thus the issue of waste prevention is dealt with separately by determination of waste reduction potentials described in Section 2.3. The hierarchy of waste treatment options is considered within the social sustainability assessment part (see Chapter 5). Within environmental assessment this aspect is included indirectly through the resources conservation aspect.

4.2 Environmental sustainability criteria and indicators

4.2.1 Selection of criteria based on Life Cycle Impact Assessment (LCIA) method

Life Cycle Impact Assessment (LCIA) is the phase in which the output of inventory analysis is further processed and interpreted in terms of environmental impacts, such as climate change or human toxicity. A number of Life Cycle Impact Assessment methods have been developed in the past years. Within the LCA-IWM project CML 2001 method based on Guinée et al. (2001) has been selected. Justification of this choice is provided in the Deliverable 3. Within this LCIA method the following baseline impact categories are recommended to be used for all LCAs (Guinée et al. 2001):

- depletion of abiotic resources
- impact of land use
    - land competition
- climate change

- stratospheric ozone depletion
- human toxicity
- ecotoxicity
    - freshwater aquatic ecotoxicity
    - marine aquatic ecotoxicity
    - terrestrial ecotoxicity
- photo-oxidant formation
- acidification
- eutrophication

An LCA study might be further extended to cover "other impact categories", e.g. loss of biodiversity, odour, noise, waste heat, casualties, depletion of biotic resources, etc (Guinée et al. 2001).

In the LCA-IWM project in order to select relevant criteria for waste management LCIA a so-called screening LCA has been performed. The purpose was to eliminate these impact categories in which impact of waste management is negligible or they do not allow any differentiation between scenarios. Selection of relevant LCIA criteria was based on full LCA modelling. LCIA including currently available and recognised impact categories was performed for two distinctive waste management scenarios for Darmstadt: highly advanced system with high level of recovery and recycling was compared to the hypothetical landfilling scenario (for details see Deliverable 3). In order to compare the magnitude of the impacts in the different categories, the characterised results have been normalised. In the normalisation step, the results are related to the overall environmental impacts in a certain region for a certain year. Thus the results can be described in e.g. Inhabitant Equivalents. In this way the following LCA impact categories have been determined as relevant for assessment of waste management scenarios:

- depletion of abiotic resources
- climate change
- human toxicity
- photo-oxidant formation
- acidification
- eutrophication.

4.2.2 Criteria based on specific targets of European waste policy

As mentioned in Section 4.1.3 compliance with targets and regulations of the European waste policy is an obligation in development and implementation of any waste management system. However, at the planning stage it is sometimes difficult to quantify performance of waste management system against specific targets. Thus the most decisive specific targets of the EU policy find consideration in the environmental assessment part of the developed tool. These include:

- packaging recovery and recycling targets,
- targets for diversion of organic waste fraction from landfilling
- collection and recovery targets of waste of electric and electronic equipment (WEEE)
- collection of hazardous waste.

Packaging recovery and recycling targets

The Directive 94/62/EC on Packaging and Packaging Waste (Packaging Directive) is concerned with specific recovery and recycling targets at the national level, thus it is not binding for municipal solid waste on a communal level (European Parliament and Council 1994). However, the national targets can only be achieved if all the administrative units within the national system contribute to the general target. The municipality itself may have targets which vary from the ones prescribed by the Packaging directive. Within this project targets of the Packaging directive will be considered as default targets for municipal waste management planning. They are compared with the achieved packaging recycling and recovery in the considered scenarios.

In the new Packaging Directive targets for the recycling and recovery of packaging waste are given (European Parliament and Council 2004). In the Assessment Tool the reference parameter is not the amount of produced packaging, but the total yearly amount of packaging waste generated by households within the municipality. Following the old Packaging Directive, the term recycling means:

> *"the reprocessing in a production process of the waste materials for the original purpose or for other purposes including organic recycling but excluding energy recovery."*

All packaging waste flows undergoing 'mechanical', 'chemical' or 'feedstock' recycling are regarded as 'recycling'. The new Packaging Directive states, that the recovery targets includes waste 'which is recovered or incinerated at waste incineration plants with energy recovery' (European Parliament and Council 2004). In this project all packaging waste flows which are fed (mono or mixed) into waste incinerators or other energy producing facilities are regarded as 'recovery'.

The targets for the recycling and recovery of packaging waste prescribed by the European Commission are shown in Table 10.

Table 10 Recycling and recovery targets for packaging waste in the EU

Target	First period EU	second period EU
Overall recovery	Min. 50%, max. 65%	Min. 60%
Overall recycling	Min. 25%, max. 45%	Min. 55%, max. 80%
Material specific recycling:		
- Glass	15%	60%
- Paper/Board	15%	60%
- Metals	15%	50%
- Plastics	15%	22,5%
- Wood	-	15%
To be attained by	June 2001	December 2008

The targets in Table 10 are valid for most of the old (before 2004) EU Member States. Greece, Ireland and Portugal should attain the targets by December 2005 and December 2011 respectively. By June 2001 only an overall recovery target of 25% is valid. For the new Member States and Pre-Accession countries later attainment dates are valid as well. The specific targets for these countries can be found in Deliverable 3.

In the output of the Assessment Tool the achieved recycling and recovery values are presented both in numbers and in a condensed overview. The targets valid for the considered country will be shown. Here a red sign for recovery indicates that the valid target for recovery was not met. A green sign indicates that the recovery target was met. A green sign for recycling means: all recycling targets met. Orange: general recycling target met, at least one material specific target was not met. Red: general recycling target was not met.

Landfill directive targets

Landfilling is ranked lowest in the waste hierarchy due to lowest potential valorisation of waste (only landfill gas utilisation). However, landfilling remains the most common waste treatment method in the European Union (ETC-WMF 2003). The Directive on landfill of waste (European Council 99/31/EC) aims at reducing the amount of biodegradable municipal waste going to landfill by the year 2016. The aim is to reach a reduction in the landfill of biodegradable waste to 35% of the total weight produced in 1995.

Limits for landfilling of biodegradable municipal waste, based on the EU Landfill directive (99/31/EC) are following:

- by 16.07.06 landfilling of 75% of the quantity produced in 1995,
- by 16.07.09 landfilling of 50% of the quantity produced in 1995,
- by 16.07.16 landfilling of 35% of the quantity produced in 1995.

Member States which transferred more than 80% of their collected waste to landfills in 1995 may postpone the attainment of the targets by a maximum of four years (achievement of targets by 2010, 2013 and 2020 apply respectively). Additionally, the Directive requires that all wastes are pre-treated prior to landfilling.

The European Topic Centre for Waste of European Environment Agency (EEA) issued a proposal for operation baseline data for respective waste streams in 1995. The baseline is subject to agreement by each member state and Eurostat (EEA 2002).

Out of the old Member States Greece, Ireland, Portugal and the United Kingdom landfilled more than 80% of biodegradable waste so for these countries later deadlines for diversion of biodegradable waste from landfills will be applied. In all the new EU Member States landfilling is a dominating waste disposal option. Thus for all these countries the delayed deadline for fulfilment of the Landfill directive targets will apply.

Targets for Waste Electrical and Electronic Equipment (WEEE)

At the moment, many electrical and electronic products are being disposed of in landfill sites. Disposal of waste electric and electronic equipment (WEEE), poses a considerable threat to the environment due to high content of toxic substances. In order to promote resources conservation and reduce the impact of these wastes on the environment a directive prescribing recovery and recycling targets for these wastes has been put into force. The directive on WEEE (European Parliament 2002) implies as a target that by the end of 2006 at least a minimum rate of separate collection on an average of four kilograms of waste electrical and electronic equipment per capita per year must be achieved. The directive also imposes recovery and recycling targets of WEEE. The recovery and recycling of WEEE is a complex issue. Until now only limited experience on this subject exists. A detailed analysis of this problem is beyond the scope of this study. Within this project environmental effects of separate collection of WEEE and recycling as opposed to the co-disposal with residual waste will be considered (see Section 9.9). No separate policy related indicator for this waste fraction is developed.

Hazardous waste management

Household waste contains hazardous waste, such as batteries, motor oils, solvents waste, light bulbs, pesticides etc. The volume of these substances is rather small, but due to their hazardous nature could undermine the treatment effects, when treated in installations designed for the non hazardous household waste only. Thus hazardous materials have to be eliminated from the household waste stream and dealt with on an individual basis. Detailed consideration of hazardous waste management is also beyond the scope of this project. Separate collection and treatment of hazardous waste is a prerequisite within any developed waste management plant. No indicator is developed to assess this aspect.

4.3 Quantification of environmental indicators

In this chapter selected criteria and respective indicator calculation methods for both LCA and waste policy-based criteria are presented. The description of selected LCA-based criteria follows CML 2001 method (Guinée et al. 2001). The criteria and indicator calculation methods will be only briefly outlined there. In practice the impacts are calculated by using characterisation factors provided by the CML 2001 method. The combination of these factors with resources consumption and emissions inventory results leads to characterised impacts of a certain system.

The characterisation factors for impact assessment of emissions within individual impact categories are provided in Appendix XVIII of Deliverable 3. For the policy related criteria the indicators calculation methods are based on respective European legislation.

Indicator: Abiotic Depletion (AbDe)

Abiotic resources are natural resources (including energy resources) such as iron ore, crude oil and others which are regarded as non living. Depletion of abiotic resources depends on ultimate reserves and rates of extraction of a given resource, which in combination provide an indication of the seriousness of the resource depletion. Ultimate reserve means the quantity of resource (as a chemical element or compound) which is ultimately available, and is estimated by multiplying the average natural concentration of the resource in the primary extraction media (e.g. the earth crust) by the mass or volume of these media (e.g. the mass of the crust) (Guinée et al. 2001). Application of this criterion within waste management assessment allows to account for positive aspects of the recovery of waste, both in form of recycling as well as energy recovery. The resources which are saved due to recycling and recovery replace abiotic resources which would have to be otherwise extracted.

The indicator **Abiotic Depletion** (AbDe) is calculated according to the following formula:

$$AbDe = \sum_{i=1}^{n} ADP_i \times m_i$$

where:

ADP_i Abiotic Depletion Potential of resource i (characterisation factor, kg Sb eq./kg)

m_i quantity of resource i extracted (kg)

Antimony (Sb) is used as a reference substance, which means that the scarcity and rate of exploitation of all other resources are related to those of antimony. Description of the origin of characterisation factor for abiotic depletion (ADP_is), according to Guinée et al. (2001) is provided in Section 7.1.1. of Deliverable 3. Default values for ADP_is for various resources are listed in the Appendix XVIII.

Indicator: Climate change (ClCh)

Climate change is defined as the impact of human emissions on the radiative forcing (i.e. heat radiation absorption) of the atmosphere. This may in turn have adverse impacts on ecosystem health, human health and material welfare. Enhanced radiative forcing causes the temperature at the earth's surface to rise. This is popularly referred to as the "greenhouse effect" (Guinée et al. 2001). Typical emissions for waste management which contribute to global warming potential include fossil carbon dioxide, dinitrogen oxide and methane (Hellweg 2003, Schwing 1999). Thus both thermal and biological waste treatment processes are relevant contributors within this criterion. Below the quantification method for climate change, based on the CML 2001 method is given.

Global Warming Potentials (GWPs) are used as characterisation factors to assess and aggregate the interventions for the impact category climate change. The overall indicator is calculated in the following way:

$$ClCh = \sum_{i=1}^{n} GWP_i \times m_i$$

where:

m_i is the mass of substance i released in kg,

GWP_i is the Global Warming Potential of the substance
ClCh is the indicator result, which is expressed in kg CO_2-equivalents.

A list of generally agreed GWP for time horizon of 100 years is provided in the Appendix XVIII of Deliverable 3.

Fossil versus biogenic carbon sources

As a general rule, only emissions of CO_2 which originate from fossil sources are accounted for within the category climate change. The CO_2 emissions which derive from biogenic sources are considered as neutral in these terms. Thus within the waste treatment modules the CO_2 emissions which stem from biological decomposition of waste are not accounted for. Similarly, within waste incineration only the proportion of carbon dioxide emissions which originate from fossil carbon sources is accounted for (e.g. CO_2 which originates from plastics burning is included, but not the one originated from burning of paper).

Carbon sequestration

An important aspect for quantification of global warming potential is sequestration of carbon. Carbon sequestration is defined as (Tzimas and Peteves 2003):

"the capture of CO_2 from its emission sources and its permanent storage"

Thus the sequestration of carbon has a decreasing effect on global warming potential: permanently stored carbon dioxide does not have any heat reflecting impact (see Section 7.2.1) anymore. Carbon sequestration (carbon from a short cycle entering a long cycle) can be considered the opposite of the burning of fossil fuels (carbon from a long cycle entering a short cycle).

In Smith et al. (2001) carbon which is fixated in such a way for a period of over one hundred years is regarded as carbon sequestration. The two occasions in which this form of carbon sequestration in a waste management system occurs are:

- landfilling of biodegradable waste
- application of compost

In both cases the carbon stems from the short carbon cycle (biodegradable waste, e.g. biowaste or paper) and is fixated. The amount of biodegradable carbon which is not decomposed after one hundred years is then assumed to be transferred to the long carbon cycle.

In the Assessment Tool project carbon fixated by the landfilling of biodegradable waste will not be regarded as sequestration. It is still strongly debatable whether or not this type of carbon sink should be included in the Kyoto Protocol (Smith et al. 2001).

Fixation of carbon into soil by compost application on the other hand, will be regarded as carbon sequestration. According to Smith et al. (2001) this type of carbon sink is considered to be included in the Kyoto Protocol.

In the calculation of the Climate Change indicator the sum of emitted carbon dioxide of fossil origin is adjusted in the following way:

$$m_{CO_2,fossil} = total\,emitted\,CO_2 - (total\,sequestered\,carbon \times \frac{44}{12})$$

Indicator: Human toxicity (HuTo)

This impact category is concerned with negative effects on human health of toxic substances emitted to the environment. The health risks of exposure in the workplace are sometimes included in LCA. However, in the CML 2001 method this kind of exposure is not considered.

Inadequate waste management practices can pose considerable threat on human health. Waste contains toxic substances which have to be managed in such a way as to minimise their penetration to the environment. Emissions from waste management with the most significant impact within this category include: heavy metals (especially hexavalent chromium, mercury, and lead (Schwing 1999), nickel and copper (Hellweg 2003), dioxins, barium and antimony (Hellweg 2003)

Indicator human toxicity is calculated according to the following formula:

$$HuTo_t = \sum_{i=1}^{n} \sum_{ecomp=1}^{k} m_{i,ecomp} \times HTP_{i,ecomp,t}$$

where:

$HTP_{i,ecomp,t}$	the Human Toxicity Potential, the characterisation factor for the human toxicity of substance i emitted to emission compartment ecomp. for the time horizon t
m_i	the emission of substance i to compartment ecomp (kg) mass of substance i released in kg,

The characterisation method adopted for this study is the one according to Huijbregts (1999 & 2000), as recommended in the CML Guide (Guinée et al. 2001). The reference substance is 1,4-dichlorobenzene. More detailed description of the indicator can be found in Section 7.3.1 of Deliverable 3.

In the provided default values in Appendix XVIII refer to the characterisation factor HTPi, time horizon - infinite, global scale, as recommended in the CML Guide (Guinée et al. 2001).

Indicator: Photo-oxidant formation (POFo)

POFo is the formation of reactive chemical compounds such as ozone by the action of sunlight on certain primary air pollutants. These reactive compounds may be injurious to human health and ecosystems and may also damage crops. The relevant areas of protection are human health, the man-made environment, the natural environment and the natural resources (Guinée et al. 2001). Photo-oxidants can be formed in the troposphere under the influence of ultraviolet light, through photochemical oxidation of Volatile Organic Compounds (VOCs) and carbon monoxide in the presence of nitrogen oxides. Ozone is considered as the most important of these oxidising agents, along with peroxyacetylnitrate (PAN). POFo is also known as summer smog or Los Angeles smog (Guinée et al. 2001).
Within waste management relevant emissions for this impact category are: Non Methane Volatile Organic Compounds (NMVOC) and methane from landfills and emissions of NOx and CO from thermal processes (Hellweg et al. 2003, Schwing 1999). Emissions of nitrogen mono-oxide have a decreasing effect on Photochemical Ozone Creation Potential (POCP), since it can react with ozone to nitrogen dioxide and oxygen, thus eliminating the effect of ozone. A photochemical ozone indicator is derived by finding conversion or reactivity factors for VOCs. This is then used to convert the inventory VOCs into ethylene equivalents. The indicator is calculated according to the following formula:

$$POFo = \sum_{i=1}^{n} POCP_i \times m_i$$

where:
$POCP_i$, the photochemical ozone creation potential of the substance i
m_i the emission of substance i

For the impact assessment within this projects a baseline characterisation method, as recommended by CML guide (Guinée et al. 2001) has been used. The characterisation factors are listed in Appendix XVIII.

Indicator: Acidification (Acid)

Acidifying pollutants have a wide variety of impacts on soil, groundwater, surface waters, living organisms and built environment. A number of man-made emissions are either acid or they are converted to acid by processes in the air. Examples of such emissions are sulphur dioxide (which becomes sulphuric acid) and nitrogen oxide (which becomes nitric acid). As for waste management the major impacts within this category arise from nitrogen oxides emissions (Hellweg 2003) from thermal processes, ammonia from biological processes (Schwing 1999) and sulphur oxide emissions from electricity production. Calculation of acidification is based on the following formula:

$$Acid = \sum_{i=1}^{n} AP_i \times m_i$$

where:

AP_i, the acidification potential of the substance i

m_i the emission of substance i

The reference substance is SO_2. Within this project a baseline method recommended by Guinée et al. (2001) applying average characterisation factors for Europe is used.

Indicator: Eutrophication (Eutr)

Eutrophication covers all potential impacts of excessively high environmental levels of macronutrients, the most important of which are nitrogen and phosphorus. Nutrient enrichment may cause an undesirable shift in species composition and surplus biomass production in both aquatic and terrestrial ecosystems. It may render surface waters unsuitable for drinking water. An increased biomass production in aquatic environment results in additional oxygen consumption for biomass decomposition (measured as Biochemichal Oxygen Demand (BOD)). Thus emissions of biodegradable matter will have the same effects as enhanced nutrient emissions, and thus BOD is also regarded as eutrophication potential (Guinée et al. 2001). Referring to the waste management the Eutrophication potential is attributed to atmospheric emissions of NOx and ammonia (Schwing 1999, Hellweg 2003), as well as phosphorus (P) and nitrogen (N) to water from biological processes. The indicator Eutr is calculated according to the following formula:

$$Eutr = \sum_{i=1}^{n} EP_i \times m_i$$

where:

EP_i, the eutrophication potential of the substance i

m_i the emission of substance i

Eutrophication is the indicator result, which is expressed in kg PO_4^{3}-equivalents.

The Eutrophication potentials according to Guinée et al. (2001) are used within this project (for characterisation factors see Appendix XVIII).

Indicator: Recycling and Recovery rates of packaging materials (RRPM)

In this chapter general recycling and recovery targets, as well as material specific recycling targets are mentioned. All plastics, glass and metals generated by households, excluding bulky waste and WEEE will be regarded as packaging waste. Non-packaging plastics, glass and metals are assumed not to be generated by households. For paper and cardboard this assumption is not valid, since a substantial share of paper and cardboard disposed of by households does not consist of packaging materials. For paper and cardboard it is assumed, that 25% consists of packaging.

The recycled amount of a single packaging waste component consists of the sum of this component in the output of the sorting facilities and for metals the output of sorted out metals of the incinerator and the MBP plants.

The total recycled amount of packaging waste consists of the sum of packaging in the output of the sorting facilities and the sorted out metals in the incinerator and MBP plants.

The total recovered amount of packaging waste consists of the sum of packaging in the output of the sorting facilities, the sorted out metals in the MBP plants and all packaging entering the incinerator or the cement kiln.

The recycling and recovery rates are determined by relating the recycled and recovered amount to the generated amount of packaging.

Indicator: Reduction of biodegradable waste landfilling (RBWL)

Although the targets of Landfill directive apply at the national level, they have to be considered at every level of waste management planning. Within this project a criterion Compliance with the Landfill directive targets will provide a waste management planner information, whether an assumed waste management scenario ensures compliance with the targets of Landfill directive at a city level and the households waste. Thus an indicator: reduction of biodegradable waste from landfilling (RBWL) is calculated in the following way:

$$RBWL = \frac{QBiodLF}{QBiod_{1995}} [\%] = \frac{\sum_{i=1}^{n} WF_i(LF) \times Biod_i}{\sum_{i=1}^{n} WF_i(1995) \times Biod_i} [\%]$$

where:

$QBiod_{1995}$ [t/year]	Quantity of Biodegradable waste generated in 1995
QBiodLF	Quantity of Biodegradable waste which is Landfilled, according to a given scenario [t/year]
WF_i (1995)	quantity of i Waste Fraction (biowaste, paper or wood, residual waste, etc.) generated in 1995 [t/year]
WF_i (LF)	quantity of i Waste Fraction which is landfilled in the considered scenario fraction [t/year]
$Biod_i$	Biologically degradable portion of i fraction of waste [%]

In order to assess the diversion rate a baseline data on biodegradable waste generation in 1995 has to be known. For that both the generation of waste by fraction as well as the rate of biodegradable material in a fraction is required. The EU Member States were obliged to report on their national methods for calculation of the rate of biodegradable material in waste to the EU Commission. Since these calculation methods could not be momentarily found for the new EU Member States, for the purpose of this project an estimate will be made based on the composition of waste in 1995 and default biodegradability level within a waste fraction. Default data on biodegradability of waste fractions is provided in **Table 18** (section 9.2).

4.4 Normalisation and aggregation of the environmental indicators

According to the principles of LCA the (i) inputs (in terms of raw materials and energy) and (ii) outputs (in terms of emissions to air, water and solid waste) are calculated for each "life stage" (see also section 4.1.3). This process is called inventory analysis. The results of the inventory are aggregated over the entire life cycle. By multiplying single emissions and resources by characterisation factors they can be attributed to the LCA-based indicators of section 4.2.1.

The total characterised values of these six indicators are expressed in Inhabitant Equivalents (IE, see Deliverable 3.1 for details). This step, the normalisation, enables a comparison between the different indicators. It should be kept in mind though, that 1 IE in e.g. the indicator Climate change does not have the identical physical meaning as 1 IE in Eutrophication. In the Assessment Tool the indicator values are shown for all 6 LCA-based criteria separately.

To enable a condensed overview, the 6 indicators are aggregated by using weighting factors and relating the total impact of a given scenario to the considered basic scenario. Because the total environmental impact as well as the impact in single categories can be either positive or negative, the condensed result is shown accordingly (negative for an environmental relief, positive for a burden). The condensed result does not have a physical meaning; it is merely a means of comparison between various scenarios. It should be noted, that weighted overall environmental impact of the basic scenario may consist of both negative and positive contributions and may therefore be close to zero. In this case the environmental condensed results of the alternative scenarios show large values (in contrary to the economic and social condensed results, where this effect does not occur).

5 Economic Assessment

5.1 Introduction

According to systems approach principles, the evaluation of the performance of any system requires the following: A **viewpoint** for the evaluation (i.e. from on whose viewpoint the system is evaluated), **objectives** which the system is expected to meet, **criteria** or **measures of effectiveness** by which the alternative actions or plans for attaining the objectives will be assessed. For a system in continuous operation, an additional requirement is the identification of the **time horizon** over which the performance will be evaluated. The performance evaluation takes place within specific system **boundaries**, which set the depth and the breadth of the analysis, thus identifying sub-systems and parallel systems.

The system in question here is the Municipal Solid Waste Management System (MSWMS). The **time horizon** of the system is two generations (about 50 – 60 years). The **viewpoint** is that of the Municipality, while the City Council is the basic decision-maker. There can of course be supplementary viewpoints, such as Regional, National, Global, depending on the chosen environmental and social indicators. For example, the impact on global warming might be of no concern to the local government (and therefore not an essential indicator), but it may be critical for the central government, especially in view of the distribution of pollution rights within each state.

The **objective** refers, in general, to a *sustainable MSWMS*, taking into consideration social, environmental and economic dimensions. This, however, might sound like a vague and immeasurable target to the local government. For a MSWMS, one might identify more specific objectives, referring to economic, environmental and social aspects, such as "satisfaction of citizens", "protection of the environment", "reduction of waste", "minimisation of social unrest", "minimisation of citizens' fees", "minimisation of cost per ton", etc. The choice of the indicators indirectly reflects on the choice for such specific objectives.

Regarding the **boundaries** of the analysis, the very essence of sustainability requires a **Life Cycle Approach**, for all three pillars. The environmental and social modules are viewed as parallel to the economic module, since they are inter-related through their common objective. In this chapter, we concentrate on the economic sustainability dimension.

As for the depth of the analysis (level of detail), one would logically expect the "same" level for all parallel modules. Yet, this is not practical for several reasons. An approximation in the order of 10 or 20% in economics has a specific meaning and it is measurable. This is not the case for social or environmental aspects. *It would be less than desirable to be very specific and "accurate" in the economic indicators and leave rather "loose" the other two piers of the analysis.* Such an approach could convey incorrect messages to the decision-makers and might defeat the very objective of the project.

According to the above, the following definitions or explanations of terms apply:

- **economic sustainability** is related (and refers) to a specific technical-organizational system, a specific time horizon and a specific decision-maker
- a system operates in an **economically sustainable manner** if it covers all its expenses and it expects to do so over the horizon of the analysis.
- if the system covers part of its expenses through subsidies, it could be considered sustainable only if there is a guarantee that these subsidies will continue to be available "forever".

Economic sustainability also implies the least expensive waste management system provided that it secures sufficient revenues to ensure

- an economically sound and continuous operation as well as
- coverage of all aftercare expenses for a period stipulated by law (certainly not less than 30 years after closure).

One should underline the difficulty in defining the same boundaries for the analysis for all three pillars of sustainability. In economics, one would rather easily identify costs and benefits related to the municipality and could separate them from those affecting neighbouring geographic areas. This is not the case for environmental impacts. The municipality is mainly interested in economic and social aspects, rather than in more vague and beyond-its-control problems like global warming (the municipality can not be held "responsible" for poor environmental performance, when it has no control over the design and the environmental performance of the collection vehicle.)

5.2 Economic sustainability criteria and indicators

5.2.1 The criteria selection process

Criteria are the rules of evaluation, the bases of comparison among *alternative forms or alternative operating mechanisms* of the system. These work as *explanations*, as *clarifications* or as *expressions* of the objectives Therefore, they must refer to the objectives. Examples of criteria: cost, quality of service, reduction in waste quantities, % of subsidies, environmental quality, equity, etc.
Indicators: These are measures of the degree to which the objectives are satisfied; in other words, they are quantitative expressions of the criteria. They also serve as *measures against preset or desired standards* (it follows that the standards should correspond to the same parameters and have the same units with the indicators).

Table 11. Example Criteria and Indicators

Criterion	Possible Indicator
1. Equity	charges [€] per capita / average income
2. Reduction of waste quantities	tons/year over a decade
3. Quality of service	frequency of collection
4. Citizen cost	actual cost [€] per HH or per capita or actual cost [€] per ton or municipal charges per ton
5. Municipality cost	€ per ton

The example criteria in Table 11 are not only economic, but also environmental (e.g. reduction of waste quantities).

The LCA dimensions: The economic analysis takes into consideration market costs incurred by the municipality, as for example the cost of a collection vehicle or of an incinerator. Whether or not these costs include all expenses in a LCA philosophy is not (and could not be) examined in this project. However, a full LCA is taken into consideration in computing the costs for the alternative waste management systems, in the sense that these costs are supposed cover all expenses from Temporary Storage to Final Disposal and Aftercare, and for a period of two generations.

Costs vs. Revenues and Credits: A central issue in our analysis is the distinction between costs incurred by the municipalities in delivering the service of solid waste management, and credits generated by recovered energy or materials or revenue from municipal rates and tipping fees. This is because the forcing factors for the costs are totally different to those for the credits and the revenues. The factors affecting MSWMS costs are different to (or, at least, do not coincide with) those affecting prices for recovered materials or tipping fees and user charges. From a management point of view, aiming at improving the system, it is necessary to separate cash outflows from revenues (cash inflows), since the "corrective" actions for the two are different. E.g., the reduction of collection costs requires different actions on the part of the municipality than the increase of charges.

In the relevant literature, one can find several cases where a different view is adapted, where costs are combined with revenues, and a net economic picture is depicted through a single indicator; such an approach would be helpful only for cities and countries with long experience in reliable waste economics accounting. It has to be remembered that this project's target users are municipalities without experience in good waste management.

5.2.2 Selected criteria and indicators

The economic sustainability criteria are grouped as follows:

- efficiency at both the sub-system level and the system level
- equity
- dependence on subsidies

Economic Efficiency

Keeping in mind the synergistic nature of the MSWMS's components (i.e. no subsystem can be effectively planned, designed and operated without due consideration of its impact on the total system and on the interactions with other subsystems), each subsystem of the MSWMS, could be identified as a "separate" entity, with its own design and operation characteristics, its own manager, its own inputs, outputs and performance evaluation. This is a usual practise in most municipalities.

In view of this approach, the objective of the economic efficiency criterion is to evaluate the impact of each subsystem on the total system's efficiency.

Possible efficiency indicator at the subsystem level: Cost per ton or per household or per person for every subsystem: Temporary Storage, Collection, Transport, Incineration, Aerobic Mechanical Biological Pretreatment, Anaerobic Mechanical Biological Pre-treatment, Landfilling, Anaerobic Digestion, Composting, Sorting, Hazardous Waste Treatment.

At the municipal level, possible indicators for the economic efficiency of the total system might be:

- cost per ton or per household or per person
- revenue from recovered material and energy
- MSWMS cost as % of Gross National Product (GNP) of the city
- diversion between revenue and expenditures for MSWMS

Equity

The purpose of this criterion is to examine the extent to which the economic burden is distributed equitably among neighbourhoods and citizens. The following indicators are suggested:

- cost per person as % of minimum wage per person
- cost per person / income per person

Dependence on Subsidies

The economic evaluation should take into consideration the financial sources for setting up and operating the system. The extent to which the municipality is self-sustainable or relying on "external" sources, i.e. on grants and subsidies, should be examined.

The suggested indicator: Subsidies or grants per person.

The initial capital investment of each subsystem and of the whole system is not an indicator of efficiency; it is, however, a significant additional factor if one examines the economic feasibility of a proposal for a new facility. All costs are transformed into equivalent annual costs, taking into consideration the time value of money (cost of capital).

For buildings, facilities and equipment, the following cost components are considered: ownership or capital recovery cost (capital opportunity cost plus depreciation), operation and maintenance, labour, salvage values, capacity reserve, etc. Overhead costs are not considered separately (except from Collection & Transport Subsystem) but are included in the above costs; for example, in labour costs, the total cost includes salary and overheads.

Environmental and social costs, in sense of estimating the economic impact (damage costs or windfall benefits) are not considered.

5.3 Quantification of economic indicators

Indicator 1: Annual Total Cost of each subsystem per ton of waste

The Annual Total Cost of each waste management subsystem per ton of waste ($AnTC_{Subsystem\,(ton)}$) is calculated (in €/ton) as follows:

$$AnTC_{Subsystem\,(ton)} = EADTC_{Subsystem\,(ton)} \frac{EADTC_{Subsystem}}{Q_{Subsystem}}$$

where

$EADTC_{Subsystem}$: Equivalent Annual Discounted Total Cost of Subsystem (in €) and

$Q_{Subsystem}$: waste quantity entering the Subsystem (in tons/year).

The Equivalent Annual Discounted Total Cost of Temporary Storage Subsystem ($EADTC_{TeSt}$) is computed as follows:

$$EADTC_{TeSt} = \sum_{i=1}^{i} \sum_{j=1}^{j} EADTC_{bins\ i\ (j)}$$

where

$EADTC_{bins\ i\ (j)} =$ $EADTPC_{bins\ i\ (j)} + EADTLC_{bins\ i\ (j)} + AMC_{bins\ i\ (j)} - EADTEC_{bins\ i\ (j)} + APC_{sacks\ i\ (j)}$

and

$EADTPC_{bins\ i\ (j)}$: Equivalent Annual Discounted Total Purchase Cost of bins of waste stream i which are used in sector j (in €),

$EADTLC_{bins\ i\ (j)}$: Equivalent Annual Discounted Total Location Cost of bins (in €),

$AMC_{bins\ i\ (j)}$: Annual Maintenance Cost of bins (in €),

$EADTEC_{bins\ i\ (j)}$: Equivalent Annual Discounted Total End-of-life Cost of bins (in €)

$APC_{sacks\ i\ (j)}$: Annual Purchase Cost of sacks of waste stream i which are used in sector j (in €).

The Equivalent Annual Discounted Total Cost of Collection and Transport Subsystem is the sum of:

- the Equivalent Annual Discounted Total Purchase Cost of Collection and Transport Vehicles (in €),
- the Annual Operating and Maintenance Cost of Collection and Transport Vehicles (in €),

- the Equivalent Annual Discounted Total End-of-life Cost of Collection and Transport Vehicles (in €),
- the Annual Personnel Cost (in €).

The Equivalent Annual Discounted Total Cost of each Treatment Subsystem includes the following:

- Equivalent Annual Discounted Capital Cost of facility
- Annual Operating and Maintenance Cost of facility
- Equivalent Annual Discounted Closure Cost of facility
- Annual Operating Cost of residual and recovered material transport vehicles
- Annual Maintenance Cost of residual and recovered material transport vehicles
- Annual Personnel (transport vehicles drivers) Cost.

The Equivalent Annual Discounted Total Cost of Disposal Subsystem includes:

- the Equivalent Annual Discounted Total Cost of landfill,
- the Annual Operating and Maintenance Cost of landfill and
- the Equivalent Annual Discounted Closure Cost of landfill.

The Initial Capital Investment of a Treatment/Disposal facility includes the following investment costs:

- land purchase
- land management and technical connection networks (access roads, water, energy supply, sewage system)
- general (non-technological) construction works
- technical installations and buildings (bunker for waste, technological buildings, storage areas for residue)
- external and internal non-technological equipment and furnishings
- technical equipment:
  - transport equipment: conveyors, loaders, trucks, tractors
  - equipment for mechanical treatment of waste: crushers, screens, balers, etc.
  - furnace system with boiler and steam utilisation system (in incineration facilities)
  - equipment for production of electrical energy
  - equipment for transport and treatment of exhaust gases
  - aerobic or anaerobic biostabilisation equipment (in aerobic or anaerobic mechanical biological pre-treatment facilities)

- waste water treatment equipment (pumps, tanks, pipes, etc.)
- other technical equipment (process control and monitoring system).

The Operating and Maintenance Cost of a Treatment/Disposal facility includes the following:

- maintenance and service of construction and equipment
- insurance
- plant management
- personnel
- electrical energy, water and heat supply
- wastewater disposal
- additional fuels
- process chemicals (lime, ammonia, adsorbent).

On the basis of an extensive literature review, a cost function was chosen for each type of facility, both for initial cost and for operating cost (Panagiotakopoulos 2002; Tsilemou and Panagiotakopoulos 2004, 2005). The development of "exact" curves, which would be useful for all possible cases in all EU countries, proved to be a futile process and an impossible one for the framework of this project.

Table 12. Suggested Cost Curves for treatment and disposal facilities (Tsilemou and Panagiotakopoulos 2004, 2005)

Type of facility	Initial Capital Investment [€][a]	Operating Cost [€/ton][a]	Range [tons/year]
Incineration[b]	$y = 5000 * x^{0,8}$	$y = 700 * x^{-0,3}$	$20000 \leq x \leq 600000$
Aer. Mech.-Biol. Pre-treatment	$y = 1500 * x^{0,8}$	$y = 4000 * x^{-0,4}$	$7500 \leq x \leq 250000$
An. Mech.-Biol. Pre-treatment	$y = 2500 * x^{0,8}$	$y = 5000 * x^{-0,4}$	$7500 \leq x \leq 250000$
An. Digestion	$y = 34500 * x^{0,55}$	$y = 17000 * x^{-0,6}$	$2500 \leq x \leq 100000$
Composting	$y = 2000 * x^{0,8}$	$y = 2000 * x^{-0,5}$	$2000 \leq x \leq 120000$
Landfill[c]	$y = 6000 * x^{0,6}$ $y = 3500 * x^{0,7}$	$y = 100 * x^{-0,3}$ $y = 150 * x^{-0,3}$	$500 \leq x \leq 60000$ $60000 \leq x \leq 1500000$

[a] Price level 2004.
[b] The incineration cost function does not include expenditures for the disposal of incineration residues.
[c] The Landfill cost function corresponds to mixed inflowing municipal waste; it does not appropriate for landfilling of residuals from other treatment facilities.

One might suggest that, given the policy decision in this project regarding the purpose, the level of practicability, and the final user, the question of accuracy for the curves in Table 12 does not arise; after all the user of the model will be warned that the curves are only rough approximations and should be used with great caution. Nevertheless, (and keeping in mind the overall target of this project which is not the exact calculation of economic, social and environmental impacts but rather the *comparative evaluation* of alternative waste management schemes at the municipal level), the degree of "accuracy" and practicality for these curves may prove to be more acceptable than the "accuracy" or practicality of information generated from the parallel environmental or social modules. Moreover, regarding the economic aspect, the local government can always compare the cost figures from Table 12 with "real" cost figure proposed by an engineering or construction company. This kind of comparison is not possible for the model predictions of social and environmental impacts.

Indicator 2: Annual Total Cost of each subsystem per household

The Annual Total Cost of each waste management subsystem per household ($AnTC_{Subsystem\ (HH)}$) is calculated (in €/HH) according to the following formula:

$$AnTC_{Subsystem\,(HH)} = EADTC_{Subsystem\,(HH)} = \frac{EADTC_{Subsystem}}{HH_{Subsystem}}$$

where

- $EADTC_{Subsystem}$ is the Equivalent Annual Discounted Total Cost of Subsystem (in €) and
- $HH_{Subsystem}$ is the number of households whose waste entering the Subsystem.

The procedure for the $EADTC_{Subsystem}$ calculation is described in § 5.3.1.

Indicator 3: Annual Total Cost of each subsystem per person

The Annual Total Cost of each waste management subsystem per person ($AnTC_{Subsystem\ (person)}$) is calculated (in €/person) according to the following formula:

$$AnTC_{Subsystem\,(person)} = EADTC_{Subsystem\,(person)} = \frac{EADTC_{Subsystem}}{P_{Subsystem}}$$

where

- $EADTC_{Subsystem}$ is the Equivalent Annual Discounted Total Cost of Subsystem (in €) and
- $P_{Subsystem}$ is the number of persons (inhabitants) whose waste entering the Subsystem.

The procedure for the $EADTC_{Subsystem}$ calculation is described in § 5.3.1.

Indicator 4: Annual Total Cost of Municipal Solid Waste Management System per ton of waste

The Annual Total Cost of MSWMS per ton of waste ($AnTC_{MSWMS(ton)}$) is calculated by dividing the Equivalent Annual Discounted Total Cost of MSWMS ($EADTC_{MSWMS}$ in €) by the studied area annual waste quantity (Q in ton) which is thrown out for collection.

$$AnTC_{MSWMS(ton)} = EADTC_{MSWMS(ton)} = \frac{EADTC_{MSWMS}}{Q}$$

Indicator 5: Annual Total Cost of Municipal Solid Waste Management System per household

The Annual Total Cost of MSWMS per household ($AnTC_{MSWMS(HH)}$) is calculated by dividing the Equivalent Annual Discounted Total Cost of MSWMS ($EADTC_{MSWMS}$ in €) by the number of studied area households (HH).

$$AnTC_{MSWMS(HH)} = EADTC_{MSWMS(HH)} = \frac{EADTC_{MSWMS}}{HH}$$

Indicator 6: Annual Total Cost of Municipal Solid Waste Management System per person

The Annual Total Cost of MSWMS per person ($AnTC_{MSWMS(person)}$) is calculated by dividing the Equivalent Annual Discounted Total Cost of MSWMS ($EADTC_{MSWMS}$ in €) by the population of the studied area (P).

$$AnTC_{MSWMS(person)} = EADTC_{MSWMS(person)} = \frac{EADTC_{MSWMS}}{P}$$

Indicator 7: Revenue from recovered material and energy

The revenue from recovered material and energy is calculated according to the following formula:

$$Reve_{MSWMS} = \sum_{s=1}^{S} Reve_{Subsystem(s)}$$

where $\mathrm{Reve}_{\mathrm{Subsystem(s)}}$ is the Annual Revenue from recovered material and/or energy of subsystems (e.g. Incineration Subsystem, Composting Subsystem).

Indicator 8: Municipal Solid Waste Cost as % of GNP of the city

This indicator measures the Municipal Solid Waste Management Cost as a percentage of the GNP (Gross National Product) of the city:

$$\frac{EADTC_{MSWMS}}{GNP_{city}} \%$$

where $\mathrm{EADTC}_{\mathrm{SWMS}}$ is the Equivalent Annual Discounted Total Cost of MSWMS (in €).

Indicator 9: Diversion between revenue and expenditures for Municipal Solid Waste Management

This indicator measures the diversion between revenue and expenditures for the MSWMS:

$$Diversion\,(revenue - expenditures) = \frac{Reve}{EADTC_{MSWMS}} \%$$

where

- Reve is the revenue from municipal solid waste taxes per year (in €) and
- $\mathrm{EADTC}_{\mathrm{MSWMS}}$ is the Equivalent Annual Discounted Total Cost of MSWMS (in €).

The revenue from municipal solid waste taxes per year is calculated by multiplying the population of the studied area by the annual taxes per person.

Indicator 10: Cost per person as % of minimum wage

This indicator measures the cost per person as a percentage of minimum wage per person:

$$\frac{EADTC_{MSWMS(person)}}{MiWa}\%$$

where

- $EADTC_{MSWMS(person)}$ is the Equivalent Annual Discounted Total Cost of MSWMS per person (in €/person) and
- MiWa is the Minimum (daily) Wage per person (in €).

Indicator 11: Cost per person / income per person

This indicator measures the cost per person as a percentage of the income per person:

$$\frac{EADTC_{MSWMS(person)}}{InPe}\%$$

where

- $EADTC_{MSWMS(person)}$ is the Equivalent Annual Discounted Total Cost of MSWMS per person (in €/person) and
- InPe is the (average annual) Income per person (in €).

Indicator 12: Subsidies or grants per person

The subsidies or grants per person are calculated according to the following formula:

$$Subsidies = \frac{EADTCC_{MSWMS} * \varphi + OpMC_{MSWMS} * \psi}{P}$$

where

- $EADTCC_{MSWMS}$ is the Equivalent Annual Discounted Total Capital Cost of MSWMS (in €),
- $OpMC_{MSWMS}$ is the Annual Operating and Maintenance Cost of MSWMS (in €),
- φ is the subsidised portion of the Equivalent Annual Discounted Total Capital Cost,
- ψ is the subsidised portion of the Annual Operating and Maintenance Cost and
- P is the population of the studied area.

6 Social Assessment

6.1 Introduction

Social Sustainability in Waste Management (SoSWM) is an integral part of sustainability in waste management, the other parts are environmental and economic sustainability. SoSWM, in broad terms, is the ethical behaviour of a waste management system towards society. In particular, this means planning and managing municipal waste responsibly towards society which has a legitimate interest in this issue and not just accomplishing legislation.

Priority aspects that should be considered in SoSWM:

- citizens' rights and obligations
- employee rights and obligations
- services suppliers' responsibilities
- government or controlling authority responsibilities
- social and environmental protection
- community involvement

These "*priorities*" are considered from three different social sustainability perspectives:

- social acceptability (MSWMS must be acceptable)
- social equity (equitable distribution of MSWMS benefits and detriments between citizens)
- social function (social benefit of MSWMS)

A list of "relevant" criteria and indicators for measuring social sustainability, classified under the next three different stages of MSWMS chain, is presented:

1.- Temporary Storage
2.- Collection System and Transport
3.- Waste Treatment

Table 13 shows the social criteria and indicators taken into account to measure the social sustainability of the MSWMS. This table shows the social perspective that measures, and the MSWMS stage that assesses each indicator.

Table 13 List of the social criteria and indicators.

Social Acceptability	Temporary Storage	Collection Transport	Treatment
1.- ODOUR	Yes	No	Yes
2.- VISUAL IMPACT	Yes	No	Yes
3.- CONVENIENCE	Yes	No	No
4.- URBAN SPACE	Yes	No	Yes
5.- PRIVATE SPACE	Yes	No	No
6.- NOISE	Yes	Yes	Yes
7.- COMPLEXITY	Yes	No	No
8.- TRAFFIC	No	Yes	Yes
9.- RISK PERCEPTION	No	No	Yes
Social Equity			
10.- DISTRIBUTION / LOCATION	Yes	No	No
11.- EMPLOYMENT QUALITY	No	Yes	Yes
Social Function			
12.- RECYCLING / DESTINATION	No	No	Yes
13.- EMPLOYMENT QUANTITY	No	Yes	Yes

6.2 Quantification of social sustainability indicators

6.2.1 Temporary Storage

Odour in Temporary Storage

Odour from waste Temporary Storage, describes potential of odour nuisance caused by a given waste Temporary Storage system to the city inhabitants. Odour nuisance from waste management storage potentially undermines social acceptance for a given waste management system.

The total odour potential of Temporary Storage is calculated as a weighted sum of respective odour potentials of individual waste fractions, according to the following formula:

$$Odour(TS) = \sum_{wf\ i=1}^{k} \left(\frac{WaFr_i}{ToWa} \times \frac{(Biod_i + WaCo_i + WaQu_i + CoFr_i + WeTe + DiPC)}{n} \right)$$

where:

wf i	separately collected waste fraction (wf), e.g. packaging waste, bio-waste, etc.
k	number of separately collected fractions [no]
$\frac{WaFr_i}{ToWa}$	weight contribution of Waste Fraction i [t/yr] to the Total Waste [t/yr] collected by the system, [%]
$Biod_i$	Biodegradability potential of waste fraction i [% DM (Dry Matter), scaled]
$WaCo_i$	Water Content of the waste fraction i [%, scaled]
$WaQu_i$	Waste Quantity of the waste fraction i per capita and year [kg/inh./yr, scaled]
CoFr	Collection Frequency of the waste fraction i [times/yr, scaled]
WeTe	Weather conditions: Temperature [°C, scaled]
DiPC	Distance Person/Container [inh./ m^2, scaled]
n	number of variables, here n = 6.

Section 6.3.1 explains how (sub-) indicators are scaled. Subsequently Odour (Temporary Storage (TS)) is normalised. For more background, see Deliverable 5.

Visual impact of Temporary Storage

Usually, visual impact is defined as a change in the appearance of the landscape as a result of development which can be positive (improvement) or negative (detraction). In our case, the waste Temporary Storage has a negative intrinsic visual impact.

This indicator measures the visual impact of the Temporary Storage (containers), taking into account the visibility, fragility and contour quality. The methodology used to calculate the visual impact indicator is an adaptation of the one proposed by Conesa (2003). First of all, the relative value of the landscape is calculated in the following way:

$$ReVL = 1{,}125 \cdot CoQu \cdot \left(\frac{PoSi}{PoDi} \cdot Acce \cdot FiViTe \right)^{1/4}$$

where:

ReVL	Relative Value of the Landscape [-]
CoQu	Contour Quality [-, scaled]
PoSi	Population Size [inh., scaled]
PoDi	Population Distance [m, scaled]
Acce	Accessibility [-, scaled]
FiViTe	Field of Vision of the Temporary Storage [-, scaled]

The visual impact indicator is calculated by means of the following equation which relates the visual impact with the relative value of the landscape:

$$ViIm = -4 \cdot 10^{-8} \cdot ReVL^4 + 9 \cdot 10^{-6} \cdot ReVL^3 - 9 \cdot 10^{-4} \cdot ReVL^2 + \\ + 3.99 \cdot 10^{-2} \cdot ReVL + 1.23 \cdot 10^{-2}$$

where:

ViIm	Visual Impact [-]

For a deeper background of the single sub-indicators contributing to the Relative Value of the Landscape, see Deliverable 5.

Convenience in Temporary Storage

The distance a citizen has to overcome to be able to dispose of his waste contributes to a large extent to his perception of the Convenience of the MSWMS. In practice this can be observed by a decrease of waste separation at higher walking distances. Also the amount of street litter increases at higher distance between dwelling and waste container.

The Convenience of a MSWMS in Temporary Storage is expressed by the average Total Walking Distance (ToWD). Assuming that all wastes considered here, are brought to the container on foot, the calculations of the total walking distance and Convenience are done as follows (with subsequent scaling and normalising of ToWD):

$$ToWD = \frac{\sum_{sector\ j=1}^{m} (NuHo_j \times \sum_{sector\ j=1}^{m} \sum_{wf\ i=1}^{k} (\frac{SuSe_j}{NBFU_{i,j}})^{1/2} \times \frac{2^{1/2}}{4} \times \frac{WaQu_{i,j}}{AAWB_i \times 10^{-3} \times NuHo_j})}{NuHo}$$

where:

ToWD	Total Walking Distance [m/hh·a]
sector j	city sector
m	number of city sectors
wf i	separately collected waste fraction (wf), e.g. packaging waste, bio-waste, etc.
k	number of separately collected fractions
$SuSe_j$	Surface per Sector j [m^2]
$NBFU_{i,j}$	Number of Bins for the waste Fraction i which are finally Used per sector j [no]
$WaQu_{i,j}$	Waste Quantity of waste fraction i per sector j [ton/yr]
$AAWB_i$	Average Amount of Waste of fraction i Brought to a TS [kg/hh/disposal]
$NuHo_j$	Number of Households per sector j [no]
NuHo	total Number of Households in the city [no]

Urban space consumption of Temporary Storage

Criterion "Urban space consumption" in Temporary Storage concerns space occupied by waste storage system (waste bins and containers) within a given city, related to the actual space availability in this city.

The indicator Urban Space Consumption in Temporary Storage (UrSC(TS)) provides a quantitative assessment of a potential space requirement for waste Temporary Storage. The indicator considers population density in the city, however it does not consider spatial city structure. Thus it does not provide detailed information on whether the free space problem really exists. This judgement is left to the indicator user, who in the final assessment can regard it as of relative importance, based on a judgement of actual space problems in the city.

The indicator expresses an area occupied by containers for waste Temporary Storage relative to the theoretical (average) city sector area available per capita (inhabitant equivalent area, scaled), using the following formula:

$$UrSC\,(TS)_{sector\ j} = \frac{SuOC_j}{AvUS_j}$$

where:

sector j	city sector

$UrSC(TS)_{sector\ j}$	Urban Space Consumption in city sector j [inh. equivalents]
$SuOC_j$	Surface Occupied by Containers in sector j [m^2]
$AvUS_j$;	Availability of Urban Space per capita in sector j [m^2/inh].

The AvUS is for each sector calculated according to the formula:

$$AvUS_{sector\ j} = \frac{SuSe_j}{PoSS_j} \qquad \left(\frac{m^2}{inhab.}\right)$$

where:

$SuSe_j$	Surface of the city Sector j [m^2]
$PoSS_j$;	Population Size of the city Sector j [inh.]

Finally the indicator is aggregated and normalised for the whole city:

$$UrSC(TS)_{norm} = \sum_{sector\ j=1}^{m} \frac{SuSe_j}{SuCi} \times \frac{UrSC(TS)_j}{4}$$

where:

$UrSC(TS)_{norm}$	Urban Space Consumption normalised
m	number of city sectors
$UrSC(TS)_j$	Urban Space Consumption in sector j [inh eq, scaled]
$SuSe_j$	Surface of the city Sector j [m^2]
SuCi	Total Surface of the City [m^2]

Private space consumption of Temporary Storage

Private space consumption is concerned with space occupied by the waste Temporary Storage inside the city inhabitants' private houses. If waste storage requires much space inside a house it may undermine acceptability of the MSWMS. This is especially true for households with small living space. Thus the indicator provides a quantitative assessment of the private space consumption by the waste Temporary Storage as a measure of the acceptance of the MSWMS.

The methodology used considers that the acceptability of the private space dedicated to the waste collection depends on four variables: number of waste fractions, floor area, household size and level of separate collection implementation in a city. Thus:

$$PrSp = \frac{NuWF + FlAr + HoSi + SeIm}{n}$$

where:

PrSp	Private Space consumption
NuWF	Number of separately collected Waste Fractions [no, scaled]
FlAr	Floor Area [m^2, scaled]
HoSi	Household Size [inh./hh, scaled]
SeIm	Separated collection Implementation level [%, scaled]
n	number of variables, here n = 4.

Subsequently PrSp is normalised.

Noise of Temporary Storage

Noise can be described as sound which affects annoyance both for human beings and animals. In general, all stages of MSWMS are afflicted with the generation of sound which can lead to noise nuisance under specific circumstances. The Noise of a MSWMS in Temporary Storage is expressed by average Container Filling Noise Potential.

The Noise caused by Temporary Storage (NoTS) is calculated according to the following method (for the whole city). For this purpose first the Container Filling Noise Potential is calculated, which is subsequently scaled.

$$NoTS_{norm} = \frac{CFNP_{scaled}}{4}$$

$$CFNP = \frac{\sum_{wf\ i=1}^{k} (NFWF_i \times NFCM_m \times NFCS_v \times \frac{WaQu_i}{AAWB_i \times 10^{-3}})}{NuHo}$$

where:

$NoTS_{norm}$	Noise caused by Temporary Storage (normalised)
wf i	separately collected waste fraction (wf)
k	number of separately collected fractions
CFNP	Container Filling Noise Potential [-]
$NFWF_i$	Noise Factor for Waste Fraction i [-]
$NFCM_m$	Noise Factor Container Material m [-]
$NFCS_v$	Noise Factor Container Size v [-]
$WaQu_i$	Waste Quantity of waste fraction i in a city [ton/yr]
$AAWB_i$	Average Amount of Waste fraction i Brought to TS per throw in [kg]
NuHo	total Number of Households in the city [no]

Complexity of Temporary Storage

The complexity of Temporary Storage depends on the waste storage/collection system itself, public awareness and recycling activities in the project area. The waste storage and collection system should be easy to understand by the citizens, so that they accept it and take part in it. This will allow a successful separation and collection of wastes and recyclables. A waste collection system can be complicated for the citizens mainly due to separate collection activities. This indicator assesses the understandability as a measure of acceptability of Temporary Storage system.

To measure the complexity in Temporary Storage, the following formula is used:

$$CoTS = \frac{SeIS + ScEd + TiSc + InWA + NuWF + Sign}{n}$$

where:

CoTS	Complexity in Temporary Storage [-]
SeIS	Separation Implemented Successfully, [-, scaled]
ScEd	School Education of inhabitants [%, scaled]
TiSc	Time Schedule [-, scaled]
InWA	number of Inhabitants per Waste Advisor [no, scaled]
NuWF	Number of Waste Fractions collected separately [no, scaled]
Sign	Signposting of container locations [-, scaled]
n	number of variables, here n = 6.

Subsequently CoTS is normalised.

Distribution and location of Temporary Storage

This criterion is concerned with equal volume distribution among inhabitants and equal walking distance for each inhabitant from a container to his living area. A waste management system requires providing equitable volume of container for each inhabitant. In reality, containers are often not distributed equally among their users and thus some may feel less privileged than the others. This indicator allows measuring the equality of distribution among all the users of a MSWMS.

Container catchment area is used to determine the equality of containers distribution. Catchment area means the area in which one container is provided for the inhabitants of this area. It is assumed that the catchment area of a container is a circle. The size of this area and the radius are calculated as following:

$$CoCA = \frac{SuSe}{NBFU} \qquad RaCA = \sqrt{\frac{CoCA}{\pi}}$$

where

CoCA	Container Catchment Area [m²/Co]
SuSe	Surface of the project area [m²]
NBFU	Number of Bins which are Finally Used in the project area [no]
RaCA	Radius of the Catchment Area [m]

Inhabitants outside the catchment area have a disadvantage with regard to their access to the Temporary Storage.

The determination of the number of households outside catchment areas can be done, if available, with a geographical information system (GIS), since in big cities with a high number of containers it would be a time-consuming work. If there is no available GIS for the project area and the manual mark of container locations on the map is not possible for the whole area, some representative areas, as areas with single family houses or with multi-family houses, etc., could be chosen.

The calculation of the share of the households outside of the container catchment areas for all households is made according to the formula:

$$HoCA_i = \frac{NHCA_i}{NuHo} \cdot 100\%$$

where:

HoCA	Percentage of Households outside of the waste fraction i Catchment Areas [%]
i	separately collected waste fraction
NHCA	Number of Households outside of the container Catchment Areas for Fraction i [no]
NuHo	Number of Households in the project area [no]

After the $HoCA_i$ values are scaled with the indicator marks for containers for fraction i, and finally the arithmetic mean is calculated for the general assessment of equity in Temporary Storage:

$$EqTS = \frac{\sum_{i=1}^{k} EqTS_i}{k}$$

where:

EqTS	Equity in Temporary Storage [-]
$EqTS_i$;	Equity in Temporary Storage for waste fraction i [-]
k	number of separately collected fractions

Subsequently EqTS is normalised.

6.2.2 Collection and Transport

Noise in Collection and Transport

The noise of Collection and Transport in a MSWMS is expressed through the increase of the mean sound level per year caused by the additional traffic through the collection of waste in relation to the existing background traffic sound level (unit: dB(A)[1]).

The model assumes that the additional traffic, generated by the considered waste transport, is distributed over the road network in the same way as the current total traffic within this network. The road net of the considered area is divided into different road types, different types of vehicles and there is also made a differentiation between day and night time (16 hours and 8 hours, respectively). There are three main input parameters which have to be known by the user. Firstly, the length of the considered

[1] dB (A) = decibel (A-weighted)

road network for each road type separately. Secondly, the mean vehicle frequency for each type of vehicle. And thirdly, the mean speed on the particular road type. Further on, it is necessary to know where the traffic for waste collecting and transport is taking place on the different types of roads (Müller-Wenk 2002).

The sound level is specific for each road type. The total of all road types per time frame (day/night) represents the increased sound level for one additional collection vehicle. The distance which causes this increase can be calculated. Thus, restricted to the vehicle category i of trucks, truck tractors, tractors, motorcycles and buses (i = 2) which is assumed to be the relevant one for this application, the collection of waste for a specific MSWMS results in two Delta Leq, differing in the variable k for day and night time. These values are added together which leads to the overall calculation result: the mean annual increase of sound level for the considered road network which is caused during waste collection:

$$DLeq_{i=2} = \left(\frac{\sum_{road\ r=1}^{x} DLeq_{r,corr./i=2/k=day}}{\sum_{road\ r=1}^{x} \left(\Delta ChNV_{r,i=2/k=day} \times LeRN_r \times 365 \times 16 \right)} \right) + \left(\frac{\sum_{road\ r=1}^{x} DLeq_{r,corr./i=2/k=night}}{\sum_{road\ r=1}^{x} \left(\Delta ChNV_{r,i=2/k=night} \times LeRN_r \times 365 \times 8 \right)} \right)$$

where:

r	road type
x	number of road categories, max x = 3 categories[2] r=1: high-performance roads (primarily conducting, without crossways, e.g. highways) r=2: main roads (primarily connecting,) r=3: side roads (primarily collecting, low speeds, foot passengers, parking on the lanes)

[2] Müller-Wenk (2002) is using four types of roads. In order to be able to implement further calculation rules as from Balzari et al. (1991) it was decided to restrict just on three types.

k	considered time frame per 24h (2 categories) k = day: 16 hours k = night: 8 hours
i	vehicle type (2 categories): i = 1: passenger cars, vans, small busses and mopeds i = 2 trucks, truck tractors, tractors, motorbikes and buses
$DLeq_{r\ corr,\ i,\ k}$	Delta noise level: Increase of sound level corrected for road r, vehicle i and timeframe k
$ChNV_{r,i,k}$	Characteristic Number of Vehicles for road r, vehicle i and timeframe k
$LeRN_r$	Length of the Road Net for road r (km)

Delta Leq is calculated both for the noise caused by collection vehicles and for the noise caused by transport vehicles in case of a transfer station. The total of these two values is used to calculate the indicator value, which is scaled and normalised.

Traffic caused by collection

Cities in all regions of Europe are concerned with heavy traffic pressure. Additional traffic caused by a MSWMS may be an important nuisance factor in a certain city. Increased traffic pressure is accompanied by various effects which are considered in other environmental, social and economic indicators, e.g. emissions, noise, odour and costs. In this criterion only the actual increase of traffic will be considered. Accompanying effects are not considered here.

Traffic Nuisance caused by Collection (TrNC) is split into two indicator aspects: Traffic Mileage (TrMi) and Traffic Blocking Potential (TrBP). Traffic Mileage is considered for the Collection and Transport phase as well as for the transport of products and residues from the treatment plants.

The Blocking Potential is only considered for the collection phase. Transport vehicles outside city boundaries are, in contrast to collection vehicles, not assumed to have any special blocking potential. Transport vehicles, as opposed to collection vehicles, do not stand idle during their transport task. Hence they are not assumed to have any special blocking potential, other than simply by increasing traffic.

Traffic Mileage: The traffic mileage is calculated according to the following formula:

$$TrMi = \frac{\sum_{wf\ i=1}^{k} TTKC_i + \sum_{wf\ i=1}^{k} TTKT_i}{NLFT/NuIC}$$

where:

TrMi	Traffic Mileage [IE (inhabitant equivalents)]
wf i	separately collected waste fraction (wf)
k	number of separately collected fractions
NLFT	National Level of Freight Transport mileage [km]
NuIC	Number of Inhabitants in the Country [no]
TTKC	Total Travelled Kilometres of Collection vehicles for waste fraction i per year [km]
TTKT	Total Travelled Kilometres of Transport vehicles for waste fraction i per year [km]

TrMi is subsequently scaled and normalised.

Traffic Blocking Potential: The traffic blocking potential is calculated according to the following formula (and subsequently scaled and normalised):

$$TrBP = \frac{\sum_{sector\ j=1,\ wf\ i=1}^{m,\ k} PiTT_{j,i} \times NuCY_{j,i}}{NuIn}$$

where:

TrBP	Traffic Blocking Potential [h/inh/year]
sector j	city sector
m	number of city sectors
wf i	separately collected waste fraction (wf), e.g. packaging waste, bio-waste, etc.
k	number of separately collected fractions
$PiTT_{j,i}$	average Pick-up Time per Trip (per sector j and waste fraction i) [h]
$NuCY_{j,i}$	Number of Collection vehicle trips per Year (per sector j and waste fraction i) [no]
NuIn	Number of Inhabitants of the city [inh]

Employment quality in Collection and Transport

As the Social Policy Agenda (CEC 2000) states: *"quality of work includes better jobs and more balanced ways of combining working life with personal life. Quality of social policy implies a high level of social protection, good social services available to all people in Europe, real opportunities for all and the guarantee of fundamental and social rights. Good employment and social policies are needed to underpin productivity and to facilitate the adaptation to change. They also will play an essential role towards the full transition to the knowledge-based economy".*

This indicator assesses the employment quality in the collection stage of the waste management system, taking into account the job characteristic and the work and wider labour market context. This indicator will be used only for an existing system. For a planned collection system it can not be applied since many aspects could not be evaluated.

The European Commission identified 10 dimensions of job quality in a Communication in 2001 (EC 2001), within two broad areas – the characteristics of the job itself; and the work and wider labour market context. The role of the indicators is to allow an assessment of how successful policies are at reaching quality in work goals across these 10 dimensions. For each of these, one or more variables have been proposed – and adopted - as a means of assessing the employment quality of the MSWMS. The employment quality is calculated as the arithmetic mean and standard deviation of the 11 variables which evaluate the 10 dimensions of employment quality (and subsequently normalised), according to the following formula:

$$EmQu = \frac{EmPhy + EmCo + PaLe + EdTra + GeRe + AcWo + EmCo +}{n}$$

$$\frac{+ FeWT + WoDL + OlWo}{n}$$

where:

EmQu	Employment Quality
EmPhy	Employees Physical effort in the collection stage [-, scaled]
EmCo	Employees Contact with wastes in the collection stage [-, scaled]
PaLe	Pay Level in the collection stage [-, scaled]
EdTra	Educational Training in the collection stage [%, scaled]

GeRe	Gender Representation in the collection stage [%, scaled]
AcWo	Accidents at Work in the collection stage [no/100.000 employees, scaled]
EmCo	Employment Contract in the collection stage [-, scaled]
FeWT	Flexible Working Time in the collection stage [-, scaled]
WoDL	Working Days Lost due to working disputes in collection stage [days/year, scaled]
OlWo	Older Workers in the collection stage [%, scaled]
n	number of variables, here n=10

EmQu is subsequently normalised.

Employment Quality in Transport

This indicator assesses the employment quality in the transport stage of the waste management system, taking into account the job characteristic and the work and wider labour market context.

The methodology used to compute this indicator is analogical to the one presented for employment quality in the collection stage.

Employment Quantity in Collection and Transport

Employment Quantity measures the total amount of direct employment in a waste management subsystem, here Collection and Transport. Direct employment means that only the number of employees contracted by a waste management subsystem is considered. Any effects, either positive or negative of indirect employment by a waste management system are excluded. Thus, effects of production of materials (containers, trucks, energy) or avoided production of substituted primary materials or energy are not accounted for.

Based on the number of vehicles and the crew size per vehicle (suggested values are given) the total number of personnnel - years for drivers and collectors is determined. The number of worked hours per employee and the average absence by illness can be entered by the user of the tool (suggested values: 1.720 hrs/yr and 3% respectively). This is also the case for the amount of employees in the overhead (all employees not being loaders and drivers): administration, planning, policy, consultants etc. (suggested: 10% of the total number of drivers and loaders).

The number of employees is, unlike other indicators, well countable. Therefore the total number of employees of a waste management scenario (consisting of employment in Collection and Transport and in Treatment) is normalised. For this purpose the total number of employees per 100.000 tons of collected and treated/disposed waste is determined by the following formula:

$$ToER = ToWa \times \frac{ToEm\,(C\,\&\,T) + ToEm\,(T)}{100.000}$$

where:

ToER	Relative Total Employment Quantity for total MSWMS [employees/100.000 t of waste]
ToWa	Total amount of Waste handled in a MSWMS [t]
ToEm (C&T)	Total Employment in the subsystem Collection and Transport of the MSWMS [employees]
ToEm (T)	aggregated Total Employment in subsystem treatment plant [employees]

ToER is subsequently scaled and normalised.

6.2.3 *Waste Treatment*

Odour in Treatment

The indicator Odour from waste treatment describes potential of odour nuisance caused by a given waste management installation to the neighbouring inhabitants.

The methodology used to quantify odour nuisance from waste treatment is based on the estimation of actual odour emission from a waste treatment plant combined with an estimation of potential scale of impact of this odour emission on the population. For the first part average odour emissions from individual waste treatment installations related to a ton of waste treated are provided. Further, the capacity of a plant is considered, in order to arrive at an estimate of total odour emission from a plant. Additionally, potential impact is being estimated based on the distance from the plant to the closest residential area.

Total Odour Emission from a single waste installation is calculated according to the following formula:

$$ToOE_t = AvOE_t \cdot WaIn_t$$

where:

$ToOE_t$	Total Odour Emission from plant t during one year [OU/yr]
$AvOE_t$	Average Odour Emission arising at specified waste treatment/disposal installation t per ton of waste treated in this plant [OU/t waste]
$WaIn_t$	quantity of Waste treated in the Installation t.

$ToOE_t$ is scaled to a value 1 to 4.

Indicator Odour (T) for a single plant is calculated taking into account the second variable, which is distance to the closest residential area (scaled as well), thus:

$$Odour(T)_t = \frac{ToOE_t + DiPI_t}{n}$$

where:

$ToOE_t$	Total Odour Emission from plant t during one year [OU/yr, scaled]
$DiPI_t$	Distance Person (closest housing estate)/Installation t [m, scaled]

The contribution of each installation to the total indicator Odour (T) in Treatment subsystem is weighted according to the weight contribution of the distance to the closest population which is effected by this plant to the sum of analogical distances for all treatment plant.
Thus the aggregated final indicator is calculated based on the following formula:

$$Odour(T) = \sum_{t=1}^{y} \left(\frac{DiPI_t}{\sum_{t=1}^{y} DiPI_t} [\%] \times ToOE_t \right)$$

where:

Odour (T)	aggregated indicator of Odour potential of subsystem Treatment [-]
t	waste treatment installation
y	number of waste treatment installations in a MSWMS
$ToOE_t$	Total Odour Emission arising at specified waste treatment/disposal installation t over one year [OU/yr, scaled]
$DiPI_t$	Distance Person (closest housing estate)/Installation t [m, scaled].

Odour (T) is subsequently normalised.

Visual impact of Treatment plants

By definition, waste is material of which people want to dispose, so it is understandable that when it comes back to their neighbourhood in the form of incinerators, landfills or large recycling plants, people will not want to accept it. One of the reasons is the negative visual impact intrinsic to the waste treatment plants. This indicator measures the visual impact of the waste treatment plants, taking into account the visibility, fragility and contour quality.

The methodology used to calculate the visual impact indicator of Treatment is the same as the one used for the visual impact of Temporary Storage (see visual impact of Temporary Storage). This methodology evaluates the visual impact of each treatment plant of the waste management system of a city. First the relative value of the landscape for each plant t is calculated in the following way:

$$ReVL_t = 1{,}125 \cdot CoQu_t \cdot \left(\frac{PoSi_t}{PoDi_t} \cdot Acce_t \cdot FiViTr_t \right)^{1/4}$$

where:

$CoQu_t$	Contour Quality around the plant t [-, norm]
$PoSi_t$	Population Size around the plant t [inh., norm]
$PoDi_t$	Population Distance to the plant t [km, norm]
$Acce_t$	Accessibility of the plant t [-, norm]
$FiViTr_t$	Field of Vision of the Treatment plant t [-, norm]

The visual impact indicator for each plant t is calculated by means of the mathematical function which relates the visual impact with the relative valour of the landscape. So, the visual impact indicator is calculated by means of the following equation:

$$ViIm_t = -4 \cdot 10^{-8} \cdot ReVL_t^{\,4} + 9 \cdot 10^{-6} \cdot ReVL_t^{\,3} - 9 \cdot 10^{-4} \cdot ReVL_t^{\,2} + \\ + 3.99 \cdot 10^{-2} \cdot ReVL_t + 1.23 \cdot 10^{-2}$$

The total visual impact of the Treatment stage of the MSWMS studied is calculated by an aggregated indicator, where the contributions of visual impacts of individual plants are weighted according to the plant capacity related to the capacity of all plants. Thus:

$$ViIm\,(T) = \sum_{t=1}^{y} \left(\frac{WaIn_t}{\sum_{t=1}^{y} WaIn_t} \% \cdot ViIm_t \right)$$

where:

ViIm(T)	aggregated indicator of Visual Impact of subsystem treatment [-]
t	waste treatment installation, e.g. a composting plant
y	number of waste treatment installations in a MSWMS
$ViIm(T)_t$	aggregated indicator of Visual Impact of plant t [-, norm]
$WaIn_t$	quantity of Waste treated in the Installation t [t/yr]

Land consumption of Treatment plant

Land consumption within waste Treatment concerns land occupation by a proposed waste treatment plant within a given city or its surroundings, related to the actual land availability of the region.

The indicator provides a quantitative assessment of a potential land consumption by waste treatment infrastructure. In the assessment two aspects are considered: the total area occupied and the type of occupation. Total land use of waste treatment is related to the population density of the city. Although the waste treatment installation does not necessarily have to be built in a city itself, through a relation to the population density in a city a sort of "footprint" of a waste treatment plant is provided. The indicator takes into account spatial structure of the region through the population density of the city. Thus it does not provide detailed information on whether the space problem exists. Analogically, to the other indicators, this judgement is left to the indicator user. Knowing the local situation, a user can decide himself how important the space problems inside the region are.

The indicator expresses land occupation by waste treatment infrastructure relative to the theoretical (average) city area available per capita (inhabitant equivalent area) together with the degree of land transformation from its previous function. For a single installation indicator is calculated according to the following formula;

$$LandUse\,(T\,)_t = \frac{\frac{SuOT_t}{AvUS} + LaTr_t + Re\,Du_t}{n}$$

where:

$SuOT_t$	Surface Occupied by Treatment plant t [m^2, scaled]
AvUS	Availability of Urban Space per capita [m^2/inh., scaled]
$LaTr_t$	Land Transformation for the location of plant t [-, scaled]
$ReDu_t$	Reclamation Duration to the state before land use for the plant t [years, scaled]
n	variable number, here n=3

The total land consumption of the treatment stage of the MSWMS is calculated by an aggregated indicator, where the contributions of land uses of individual plants are weighted according to the surface occupied by this plant related to the summary surface occupied by all plants in the MSWMS, thus:

$$LandUse\,(T) = \sum_{t=1}^{y}\left(\frac{SuOT_t}{\sum_{t=1}^{y} SuOT_t}\% \cdot LandUse(T\,)_t\right)$$

where:

Land Use (T)	aggregated indicator Land Use of subsystem treatment [-]
t	waste treatment installation, e.g. a composting plant
y	number of waste treatment installations in a MSWMS
Land Use $(T)_t$	Land Use of treatment plant t [-, scaled]
$SUOT_t$	Surface Occupied by Treatment plant t [m^2]

Land Use (T) is subsequently normalised.

Traffic caused by Treatment

Traffic generated by Treatment considers the transport of secondary materials or secondary waste streams leaving the treatment plants.

Traffic Nuisance caused by Transport of treatment Products (TNTP) is expressed in only one indicator aspect: Traffic Mileage (TrMi). The Traffic Blocking Potential is not considered here. Traffic Nuisance expresses the nuisance which is generated by the extra traffic caused by a MSWMS, by extra driven kilometres (a gathering of a range of secondary effects).

The indicator aspects are calculated according to the following proposed method:

$$TrMi(T)_t = \frac{\sum_{pr\ p=1}^{z} \frac{TrDS_p \times PrFl_p \times ReTF_p}{NTLS_p}}{NLFT / NuIC}$$

where:

$TrMi\ (T)_t$	Traffic Mileage for plant t [IE (inhabitant equivalents)]
NLFT	National Level of Freight Transport mileage [km]
NuIC	Number of inhabitants in the country [no]
$TrDS_p$	Transport Distance of Secondary product/waste p [km]
pr_p	secondary product or secondary waste fraction p
PrFl	Product Flow of secondary product/waste p [ton/yr]
NTLS	Net Truck Load for Secondary product/waste p [ton]
ReTF	Return Tour Factor (2 for an empty return freight)

For a single plant the indicator is normalised (see Section 6.3.1). The aggregated indicator for the subsystem Treatment is calculated as the sum of Traffic Milage indicators of all plants, thus:

$$TrMi(T) = \sum_{t=1}^{y} TrMi(T)_t$$

where:

TrMi (T)	aggregated Traffic Milage [IE]
t	waste treatment installation, e.g. a composting plant
y	number of waste treatment installations in a MSWMS
$TrMi\ (T)_t$	Traffic Milage for plant t [IE]

TrMi (T) is subsequently scaled and normalised.

Risk perception in Treatment

Risk perception refers to an individual's cognitive process of attribution of meaning. The term *Risk Perception* describes how people react to a given risk. These reactions have a number of dimensions and are not simply reactions to any physical hazard itself.

The indicator quantifies the population risk perception towards waste treatment plants by means of a questionnaire survey of the population. The survey involves seven key factors affecting public risk perception: trust, voluntary versus involuntary, control, benefit/reward, understanding, gender and catastrophic potential.

The extensive research in the field of qualitative assessment of risk perception has identified a number of key factors affecting public perception of risk and thereby public acceptance of a given level of risk. The key factors relevant in risk perception are briefly reviewed below (OECD 2002): Trust, Voluntary versus involuntary, Control (lack of personal control vs. feeling of control over a situation), Benefit/Reward, Understanding, Gender and Catastrophic potential

The survey questionnaire has twenty-seven questions. Each question is related to a key point of the risk perception which has been commented before.

The questions are designed as multichotomous or by Likert-scale. In the multichotomous questions respondents are asked to choose the alternative that most closely corresponds to the position on the subject. Likert-scale questions are used for perception related questions for gathering information on attitudinal behaviours of people. Respondents are asked to complete a five point Likert scale on each question or proposition.

The Sampling Procedure is a probable sample, a chosen subset of the population of interest which ensures a representative cross section by giving every element in the population a chance, nonzero chance of being selected. This is done by random selecting the sample from that population.

Employment Quality in the Treatment

This indicator assesses the employment quality in the Treatment stage of the waste management system, taking into account the job characteristics and the work and wider labour market context.

The methodology used to calculate the employment quality in the waste treatment plants is the one proposed to calculate the employment quality in the Collection and Transport stage of the waste management system.

Final destination

This indicator measures the social function of the used waste management option taking into consideration the recovery rate.

The indicator determines an overall level of recovery in the studied region, by evaluating three variables: energy recovery rate, recycling rate and composting rate. Each individual variable is computed as a percentage of recovered (energetically, recycled, or composted) waste in relation to total waste generated in the region. These three variables are normalised according to the achievable recovery rates in the old and new European states (respectively for a given region). The final indicator is calculated in the following way:

$$ReRa = \frac{RcRa + CoRa + EnRe}{n}$$

where:

ReRa	Recovery Rate in the project area [-]
RcRa	Recycling Rate [%, scaled]
CoRa	Composting Rate [%, scaled]
EnRe	percentage of waste which is used to Energy Recovery [%, scaled]
n	number of variables, here n=3.

ReRa is subsequently normalised.

Employment Quantity in the Treatment

Employment Quantity is the total amount of direct employment in the Treatment subsystem in a waste management system.

The methodology is the same which is used in the indicator of the employment quantity in the Collection and Transport stage of the MSWMS. The user of the tool can enter the total number of employees of a certain treatment plant. This information will mostly not be available to the user, since the considered scenarios are not actually implemented in the city. For this reason the tool provides with a default value for the relative number of employees ($REQu_t$). The provided default value consists of a mathematical relation between the plant capacity and the relative number of employees (in employees per 100.000 t annual capacity). The used relationships can be found in Deliverable 5. The absolute number of employees for the considered treatment plant follows from the following formula and depends on the relative number of employees and the input of the plant (only the input from the considered city):

$$ToEm(T)_t = \frac{InWI_t}{100{,}000} \times REQu_t$$

where:

$ToEm(T)_t$	Total Employment in a considered treatment plant t [employees]
$REQu_t$	Relative Employment Quantity in a considered treatment plant t [default employees/100.000 t input]
$InWI_t$	Installation Waste Input of a treatment plant t [t/year]

If actual absolute number of employees is known for the considered plant, the suggested value can be overruled by the user. The values for $ToEm(T)_t$ of the various plants lead to the total number of employees in the Treatment subsystem of the MSWMS:

$$ToEm(T) = \sum_{t=1}^{y} ToEm(T)_t$$

where:

ToEm (T)	aggregated Total Employment in subsystem treatment plant [employees]
t	waste treatment installation

y	number of waste treatment installations in a MSWMS
$ToEm(T)_t$	Total employment in a single treatment plant t [employees]

The total number of employees in the Treatment subsystem is aggregated with the employment in the Collection and Transport subsystem (see Section 6.2.2).

6.3 Aggregation of social indicators

The indicator mark indicates the level of social acceptability, equity and function of the indicator evaluated in the MSWMS. It depends on the independent variables values. The indicator mark is the result of the indicator independent variables taking a certain value at the same time.

6.3.1 Indicator calculation methodology

1.- *Define independent variable values:* possible values (or situations) are assigned to the independent variable, they are classified in five levels from the best situation (the one socially acceptable, equity and function) to the worst (the socially unacceptable, inequity and functionless). The user should choose the most feasible variable value to his MSWMS.

2.- *Scaling:* A mark (ranging from 0 to 4) is assigned to each independent variable situation. This mark is defined by the social sustainable level assigned before; 0 means the best situation from a social sustainability point of view and 4 the worst situation. These marks are assigned internally by the tool. The ranges of the marks that are used in the tool are described in detail in Deliverable 5.

3.- *Indicator calculation:* The indicator is calculated by a function which combines the features which are considered relevant for the indicator. These features have been called "independent variables". In order to achieve all the assumptions, conditions and physical sense as far as possible the indicator is calculated with the scale variable value. These calculations are done internally by the tool:

$$Indicator = \frac{v_1 + v_2 + v_3 + \cdots\cdots\cdots + v_{n-1} + v_n}{n}$$

4.- *Indicator normalisation:* The indicator result is normalised between 0 and 1. The value 0 represents the best situation and 1 the worst situation. In this way, it is possible to reduce/condense information, measure and compare the components of single or similar systems with one indicator.

$$Indicator\_normalised = \frac{Indicator}{4}$$

6.3.2 Indicator weighting

Weighting is the process of converting indicator results of different impact categories into scores by using numerical factors based on values. Weighting may include aggregation of the weighted results into an overall score. This is a subjective stage based on value judgements and is not scientific. ISO 14042 cautions that different individuals or organisations may have different preferences or values. Therefore, different parties are each likely to obtain different weighting results based on the same indicator results.

The user should assign a weight (score) to each indicator in the different waste management system stages. This weight is assigned depending on the social acceptability, equity and function of the indicator by the population. Because of that, the end-user should base his decision on the population's priorities and/or social complaining. These weighting factors should reflect the population's feeling towards the social themes expressed by the indicators. A lower weight means that the theme is not very important while a higher weight indicates the more important themes. The aggregated social sustainability results are obtained by a weighted average of the weighting factors and the single indicator scores.

7 Temporary Storage

Temporary Storage is the point where household waste leaves the household and enters the waste management system. The waste is temporarily stored in bins, containers and sacks prior to collection. Figure 8 shows the throughput and the impacts of Temporary Storage.

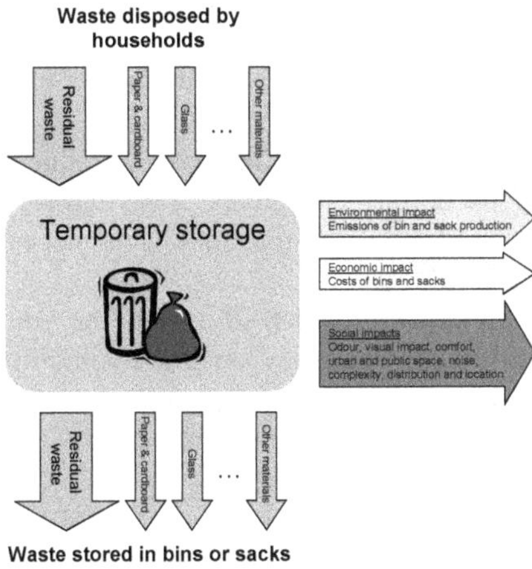

Figure 8. Temporary Storage.

The input is based on the source separated waste streams which have been estimated based on the implementation plans (see Section 2.3). Output are the different waste streams which are stored in sacks, bins or containers – either in kerbside or bring system.

The direct impacts of this stage contain the

- environmental emissions due to emissions of bin and sack production,
- the total economic costs of this stage as well as
- a set of social impacts with regard to the social acceptability and social equity of the waste management system.

Out of these impacts the encouragement of social acceptability plays an important role. It has to be noted that the Temporary Storage as contact point between waste generators and waste management system has to be carefully managed. The households need to have the solid waste collected with a minimum of inconvenience (in terms of e.g. odour, noise, complexity), while the collector needs to receive the waste in a form and quantity (e.g. recyclables) compatible with the planned treatment methods. Waste management systems which fail to achieve balance in this relationship are unlikely to succeed (McDougall et al. 2001).

Concerning the design of Temporary Storage municipalities have the following relevant decision options:

- **Volume of used containers**: Depending on the collection system different sizes of bins and containers could be appropriate. These can vary from single-household bins (starting from 80 l within LCA-IWM) to multi-household containers (with a maximum volume of 5000 l).
- **Collection frequency**: The collection frequency determines the number of used bins and sacks. The higher the collection frequency, the lower will be the number of necessary collection repositories.

Table 14 shows an example which illustrates the impact of the selected bin volume and collection frequency on the availability of the bins. The higher the bin volume and collection frequency, the lower is the distance to the next bin, thus the convenience for the citizens.

Table 14. Impact of bin size and collection frequency on the bins density.

Assumed collection quantity and settlement density (fictitious example)				
Collected waste fraction:	Paper and cardboard			
Bulk density of fraction:	140 kg/m³			
Collection quantity:	20 kg/cap/yr			
Settlement density in the sector:	3.000 inhabitants/km²			
Selected design options				
Bin size [litres]	240	770	2200	5000
Collection frequency [#/year]	12 (monthly)	26 (biweekly)	52 (weekly)	104 (2/week)
Results				
Bin density [inh./bin]	20	140	800	3640
Catchment area [ha/bin]	0.67	4.67	26.7	121
Max. distance to a bin [m][a]	90	240	580	1240
Collection point	Kerbside	Street-side	Central	Central

[a] Assumed relation between the way and air-line distance of 2.0.

7.1 Process description

Dimensioning necessary bins and sacks for an actual or future year
For the calculation of the number of needed bins and containers all waste streams are treated equally. The same method of calculation is applied both for already implemented separately collected waste flows and planned flows. For the already existing collection, although the actual numbers are known, a hypothetical number of bins and containers is determined. In this way any favouring of either the existing or planned collection flows is prevented. A consequence of this approach is that the modelled number of bins and containers may deviate from actually implemented or detailed planned collection schemes.

Selecting the types of used bins and sacks
Concerning the main collection streams within the waste management system (residual waste, paper and cardboard, glass, metals, plastics and composites, packaging material and bio-waste) the user can select up to three different types for the Temporary Storage, such with

- sack collection (mainly kerbside collection),
- collection with bins with less than 500 litres volume (mainly kerbside collection) and
- collection with containers with more than 500 litres volume (street-side, at material banks or at central collection sites).

Table 15 shows the volume and materials which can be selected. For each of these three collection modes (sacks, small and big-sized bins) one of the listed types which is exclusively (or even predominantly) in use in the City can be selected by the user.

The remaining four waste types (garden waste, hazardous waste, WEEE and bulky waste) are usually not collected with sacks or small bins which is due to their bulky character (garden waste, WEEE and bulky waste) or due to the importance of central collection in case of the hazardous waste (McDougall et al. 2001). Therefore a lot of different container types (e.g. sometimes no container in case of separate and infrequent collection of bulky or garden waste) are in use. As the main material consumption and bin costs in MSWMS come from small bins, it was acceptable to neglect the use of these containers.

Table 15. Possible volume and materials of sacks and bins.

Sacks		Bins with volume below 500 l		Bins with volume above 500 l	
Volume [l]	Material	Volume [l]	Material	Volume [l]	Material
60	PE	80	HDPE/Steel	660	HDPE
80	PE	120	HDPE/Steel	770	HDPE/Steel
110	PE	240	HDPE	1100	HDPE/Steel
				2500	HDPE/Steel
				3200	Steel
				5000	Steel

PE ... polyethylene
HDPE ... High density polyethylene
Steel ... Galvanized steel plate

Estimating the number of necessary bins and sacks on sector level
The estimation of the necessary number of bins and sacks (which are now defined, see the last section) depends on the following parameters:

- percentage of inhabitants using 1. sacks, 2. small bins (volume below 500 l) or 3. big bins (volume higher than 500 l) for each fraction
- collection frequency for each type of Temporary Storage, sector and fraction (# per year)
- average filling rate of bins: This correction factor considers the fact which bins are not totally filled at every collection cycle (e.g. due to seasonal variation)
- waste density per fraction (constant parameters)

Based on these inputs and on the separated collected waste quantities (from Chapter 2.3), the number of necessary bins can be calculated.

The detailed description of calculations and model parameters of the Temporary Storage can be found in the Section 8.4. of Deliverable 3.

8 Collection and Transport

Collection and Transport includes the

- collection of unseparated and separated solid waste and recyclables in an urban area and the
- transportation of the collected waste and recyclables to processing and disposal facilities.

Figure 9 shows the scheme of this process. Input contains the material in bins and sacks at the time of collection which will be transferred to the provided facility.

The direct impacts of this process contain the

- traffic emissions deriving from the necessary transports,
- the economic costs covering the costs for personnel, truck fleet (purchase and maintenance costs), fuel cost etc. and
- some social impacts with regard to the social acceptability, equity and function of the Collection and Transportation management in the waste management system.

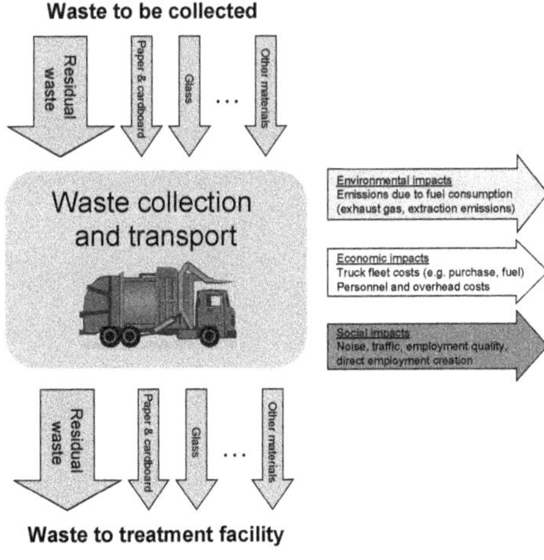

Figure 9. Waste Collection and Transport.

Previous experiences point out the high relevance of collection costs which can reach a percentage of 50 to 80 percent (Bilitewski et al. 1996, Tchobanoglous et al. 1993) of the total MSWMS costs. Also social impacts, e.g. employment quality and job creation, are noted as relevant criteria of the design of a logistical system (Bilitewski et al. 1996, Tchobanoglous et al. 1993).

8.1 Process description

The model applied in this tool was created to fulfil the following goals with regard to necessary simplification restrictions, as the realistic estimation of the necessary

- transport distances (as basis for fuel consumption and costs and social impacts (e.g. noise)),
- expenditure of time for Collection and Transport (for personnel costs) and finally the
- required truck fleet capacity (e.g. for purchase cost calculation).

Waste Collection and Transport routes
The transfer from the Temporary Storage to the Treatment Facility or Landfill consists of two processes:

- **collection** covers the emptying of bins or/and sacks in the settlement area
- **transport** refers to the haulage of the collected waste to the facility or treatment plant.

Transportation by road has been assumed for both processes. The literature confirms the subordinate role of rail or ship transportation in existing waste management systems (Bilitewski et al. 1996, Tchobanoglous et al. 1993, McDougall et al. 2001).

Two general cases have been assumed concerning the logistics management of a defined waste fraction:

1. Collection and reloading at a transfer station prior to transport (Case 1)
2. Collection and direct transport to the facility or landfill (Case 2)

Figure 10 shows the Collection and Transport scheme with an existing transfer station (Case 1) consisting of the collection (Case 1a) and the transport (Case 1b) with different vehicles. In case of a short distance from the collection area to a facility (even inside the area) the combined Collection and Transport with one vehicle type has been offered as alternative design option for each waste fraction (Case 2). Table 16 shows an overview of all sub-processes for these two options on a working day.

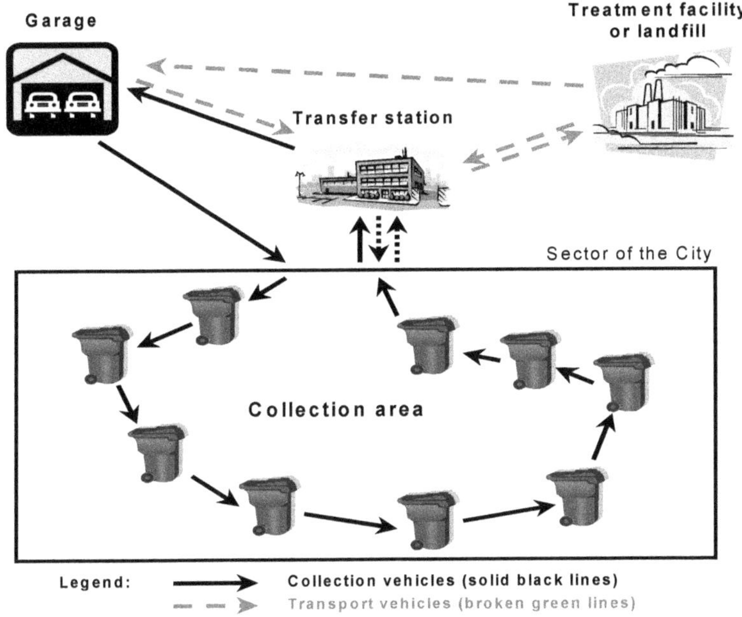

Figure 10. Collection and Transport scheme with transfer station (Case 1).

Table 16. Collection and Transport sub-processes on a working day.

Collection (Case 1a)		Transportation (Case 1b)		Collection and Transportation (Case 2)	
#	Process	#	Process	#	Process
1	Garage – sector	1	Garage – TS	1	Garage – sector
2	Collection in sector	2	Loading at TS	2	Collection in sector
3	Sector – TS	3	TS – Facility	3	Sector – Facility
4	Unloading at TS	4	Unloading at facility	4	Unloading at facility
5	TS – sector[a]	5	Facility – TS[a]	5	Facility – sector[a]
6	Collection in sector[a]	6	Loading at TS[a]	6	Collection in sector[a]
7	Sector – TS[a]	7	TS – Facility[a]	7	Sector – Facility[a]
8	Unloading at TS[a]	8	Unloading at facility[a]	8	Unloading at Facility[a]
9	TS - Garage	9	Facility - Garage	9	Facility - Garage

TS ... Transfer station

[a]For every additional collection or transport trip on a working day.

The user can specify the applied Collection and Transport scheme in the following way:

For each sector:

- average distance from the garage to the first pick-up in a defined sector (enter zero, if garage is in the sector)
- average distance from the transfer station to the first pick-up in a defined sector (enter zero, if no transfer station is existing for this sector)
- average distance from the transfer station to the garage (enter zero, if no transfer station is existing for this sector)

For each waste fraction and sector:

- specification, if a transfer station is used (or even existing) for this fraction or not
- average distance from the transfer station to the designated facility (Case 1)
- average distance from the sector to the designated facility (Case 2)

More details on the Collection and Transport subsystem can be found in section 8.5 of Deliverable 3.

9 Treatment

9.1 Introduction

Within the developed Assessment Tool the following waste treatment processes have been modelled:

- composting of separately collected organic waste
- digestion of separately collected organic waste
- aerobic mechanical-biological pre-treatment (MBP) of mixed/ residual waste
- anaerobic MBP of mixed/residual waste
- incineration with energy recovery of mixed/residual waste
- landfilling of mixed/residual waste
- recycling of separately collected materials: paper and cardboard, glass, metals, plastics, packaging waste, mixed dry recyclables (MDR) and waste electric and electronic equipment (WEEE).

The technologies selected for modelling are the ones most commonly used in modern waste management systems in Europe. They are considered as state-of-the-art, but already broadly verified treatment methods.

9.2 Chemical properties of waste

Input data on chemical properties of waste is a basis for modelling of mass balances of waste treatment processes. Chemical composition determines important features of waste such as e.g. biodegradability rate or calorific value. Contaminants content in waste (heavy metals, chloride, fluoride, etc.) renders the quality of products derived from the waste. In the following sections default values for waste characteristics are provided.

9.2.1 Separately collected organic waste

Default characteristic of separately collected organic waste, divided to bio and garden waste, is provided in table 17.

Table 17. Default characteristic of separately collected organic waste (Vogt et al. 2002, Fricke et al. 2002c)

Parameter	Unit	Biowaste	Garden waste
Dry matter (DM)	%	45,0	43,0
ODM	%DM	87,0	84,0
bioODM	%ODM	100	100
C	%ODM	51,9	49,8
Nitrogen (total)	%DM	1,7	1,2
Phosphorus (total)	%DM	0,4	0,5
Potassium (total)	%DM	0,9	1,5
Magnesium (total)	%DM	0,8	0,5
Calcium (total)	%DM	2,2	4,4
Cadmium	mg/kg DM	0,1	0,3
Chromium	mg/kg DM	1,8	4,6
Copper	mg/kg DM	9,2	0,1
Mercury	mg/kg DM	0,004	0,2
Nickel	mg/kg DM	1,3	3,7
Lead	mg/kg DM	2,6	4,8
Zinc	mg/kg DM	30,6	60,0

9.2.2 Mixed and residual waste

Default characteristic of residual waste fractions is provided in Table 18. Table 19 contains default values on heavy metal contents of individual fractions of residual waste.

Table 18. Default characteristic of residual waste. based on Rotter (2004). Fricke et al. (2002c) and Dehoust et al. (2002).

	DM	ODM	biol. ODM	C	biogene C	H	O	N	Cl	S
Waste material	[%]	[%DM]	[%ODM]	[%ODM]	[%C]	[%ODM]	[%ODM]	[%ODM]	[%ODM]	[%ODM]
Paper	62	87	98	49	99	6,4	44	0,2	0,3	0,2
Glass	95	0	0	47	98	10,0	40	3,0	0,0	0,0
Iron	90	0	0	48	98	6,3	44	0,5	0,7	0,1
Aluminium	90	0	0	48	98	6,3	44	0,5	0,7	0,1
Copper	90	0	0	48	98					
Plastics	85	95	5	83	5	13,3	4	0,1	0,1	0,0
Packaging composites	75	91	78	59	60	6,7	39	2,7	0,7	0,5
Composites	85	80	58	58	20	6,7	39	2,7	0,7	0,5
Biowaste	45	87	100	51	100	6,2	44	0,5	0,1	0,1
Garden waste	43	84	100	50	100	7,9	32	0,0	0,7	0,0
Wood	80	90	50	49	100	7,6	33	0,5	1,5	0,1
Diapers	50	50	25	57	90	7,7	31	3,6	0,8	0,3
Inerts	90	0	0	48	98	6,3	44	0,5	0,7	0,1
Textiles	70	85	60	51	65	6,9	37	4,3	0,4	0,4
Leather	70	85	50	47	90	6,4	40	2,0	0,7	0,3
Medium corn	56	49	88	47	65	6,5	40	2,5	0,5	0,6
Fine corn	66	39	88	49	100	7,0	33	13,0	0,0	0,0
Hazardous	75	50	25	70	0	9,9	19	0,4	0,7	0,1

DM = Dry Matter
ODM = Organic Dry Matter
biol. ODM = biologically degradable Organic Dry Matter

Table 19. Heavy metals content in household waste fractions (Morf and Brunner 1999)

Waste material	As	Cd	Cr	Cu	Hg	Ni	Pb	Zn
				[mg/kg DM]				
Paper	5,0	0,7	9,8	45	0,2	6,8	23	295
Glass	0,0	0,0	1,3	0	0,0	0,0	329	82
Iron	20,0	21,0	156	265	-	68	582	507
Aluminium	20,0	21,0	156	265	-	68	582	507
Copper	20,0	21,0	156	1E+06	-	68	582	507
Plastics	5,0	66,0	28,6	60,4	0,2	4,3	50	627
Packaging composites	5,0	1,0	36	68,0	0,2	7,4	30	388
Composites	5,0	1,0	7,3	37,5	0,2	9,0	14	90
Biowaste[a]	30,0	1,0	55	153,0	0,5	28	90	500
Garden waste[a]	30,0	1,0	55	153,0	0,5	28	90	500
Wood	5,0	0,4	5,5	17,9	0,1	3,8	21	158
Diapers	5,0	0,5	27	23,2	0,2	11	10	313
Inerts	10,0	0,5	80	35,0	0,1	45	50	70
Textiles	5,0	1,0	16,8	55,0	0,1	7,3	35	170
Leather	5,0	3,0	900	43,0	0,1	5,1	112	4.438
Medium corn[b]	30,0	1,0	55	153,0	0,5	28	90	500
Fine corn	35,0	2,0	75	715,0	1,1	44	190	780
Hazardous	12,0	53,0	17,5	1690	127	347	10.800	106.000

[a]only data for organic waste available, thus no differentiation was made between bio and garden waste
[b] no data available, thus the same concentrations in organic waste were assumed

9.3 Composting of organic waste

Composting is a controlled process of biological decomposition of organic material in the presence of air to form a humus-like material (EEA 2003a). Feedstock is separately collected organic waste fraction from households. Both biowaste and garden waste are suitable for composting. Most common technologies applied in centralised composting plants include: static pile (windrow), aerated static pile (aerated windrow), agitated bed, rotating drum and composting box. The composting technology chosen for modelling within the LCA-IWM tool is box composting with subsequent maturation in enclosed windrows. Advantage of a fully encapsulated system is the reduction of odour which is the main nuisance caused by a composting process. Selection of a modular system, such as box technology allows adjusting the plant capacity to the current waste input.

9.3.1 Process description

The LCA-IWM module "Composting plant" describes composting process in a two-step technology: (i) intensive composting in composting boxes and (ii) maturing process in windrows in a composting hall. Prior to composting, the separately collected organic waste undergoes a mechanical pre-treatment, in which contaminants are sorted out. Within box composting fresh compost is produced. Full automatisation of the intensive composting shortens the time of fresh compost production to 10 – 14 days. The fresh compost is further treated in the maturing step, which takes another app. 8 weeks. After the composting period of altogether 10 weeks, the product can be used as a high quality compost (maturity grade IV or V, according to the German composting grading). Both steps of composting process cause air and water emissions. In an enclosed composting plant air emissions can be collected and purified before discharge into the atmosphere. In the developed model the user has an option to select purification of flue gas in a biofilter.

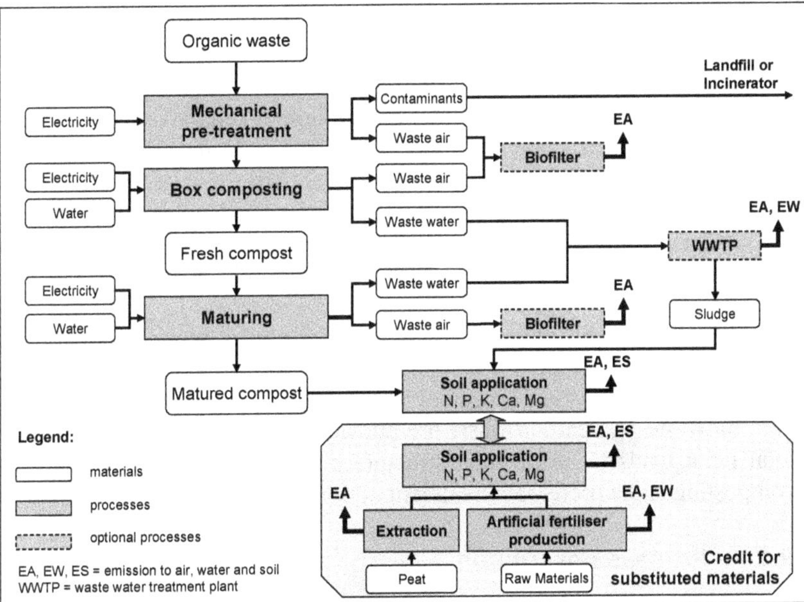

Figure 11. Schedule of the main material flows in the modelled Composting plant

Also waste water is collected and an option of treatment in a waste water treatment plant (WWTP) can be chosen. Direct discharge of emissions into the atmosphere and effluent receiver is not recommended, but the model provides such option in order to allow modelling of already existing plants. The produced compost as well as the sludge from the WWTP is applied on agricultural land by landspreading. A credit is accounted for the nutrient content of the compost. It is assumed that they substitute the use of artificial fertilisers. In Figure 11 main material flows of the composting plant which were modelled in the Composting plant module are provided.

Parameters applied for modelling of the composting process are described in the Section 8.7 of Deliverable 3.

9.4 Digestion of organic waste

Digestion is a process in which biodegradable substances are decomposed by bacteria. The degradation process is performed under anaerobic conditions and leads to the production of biogas. The produced biogas is a mixture of mainly methane and carbon dioxide. Digestion as a waste treatment method is applied with a variety of substrates, such as manure, slaughterhouse waste, waste water treatment sludge and biowaste. Also combinations of substrates, the so-called co-digestion or co-fermentation is widely applied. In this project only the mono-digestion of separately collected organic waste (biowaste and/or garden waste) is considered. Digestion processes can be divided into wet or dry, thermophilic or mesophilic and 1-stage or 2-stage processes. For this project a thermophilic dry 1-stage process has been modelled. This is the most common process type for the digestion of biowaste in Germany (Fricke et al. 2002a, Kern 1999, Vogt et al. 2002).

Although materials of good structure, which garden waste mostly is, are principally more suitable for composting, they can be digested as well, especially in dry digestion processes (FNR 2004). In the digestion module both biowaste and garden waste are allowed waste inputs. If however the total input tends to be of good structure material (wood like, celluloses) composting is the preferable treatment option.

9.4.1 Process description

Prior to digestion, the separately collected biowaste undergoes a mechanical pre-treatment, in which contaminants are sorted out. In the actual digestion phase biogas, wastewater and a digestion residue are produced. The biogas is combusted in an engine to produce electricity and heat. The residue is further treated in the aerobic maturation process stage, producing compost. Waste water is treated in a waste water treatment plant (WWTP).

The produced compost as well as the sludge from the WWTP is applied on agricultural land. The nutrients contained in the compost substitute artificial fertilisers. The production and application of these are accounted as credits.

In Figure 12 a schematic overview of the Digestion plant that was modelled in the LCA-IWM tool can be seen.

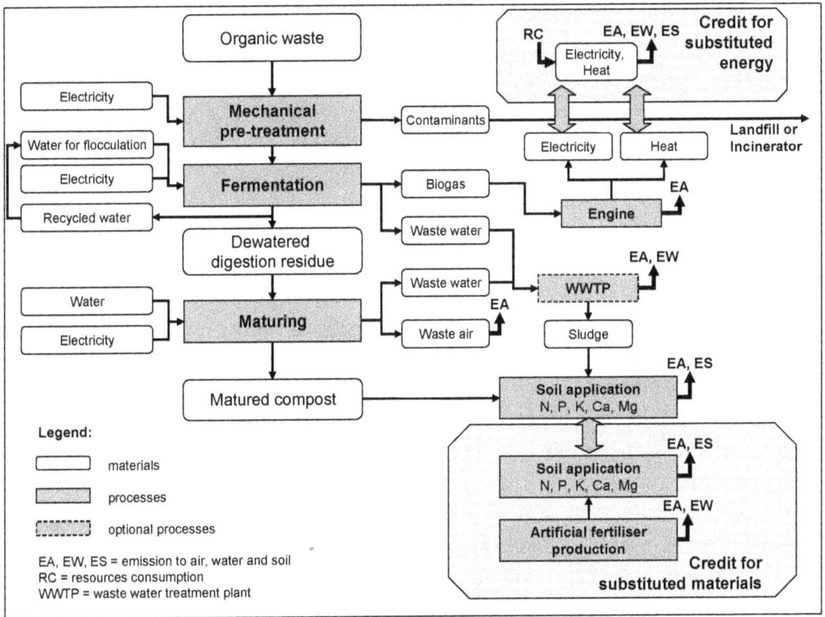

Figure 12. Schedule of the main material flows in the modelled thermophilic dry 1-stage Digestion plant

The modelling of the Digestion plant is mainly based on data of German plants. However, the outcomes depend strongly on the input data which can be provided by the user. Parameters used for modelling of digestion process are provided in Section 8.6 of Deliverable 3.

9.5 Aerobic Mechanical-Biological Pre-Treatment

Mechanical-biological pre-treatment (MBP) is a method alternative to the incineration of pre-treatment of mixed or residual waste prior to landfilling. The concept of the biological treatment in aerobic MBP plant is similar to composting. The aim of MBP is to minimise the environmental impacts of landfilling and to gain value from waste through metals and energy recovery. The main MBP technologies are based on either "splitting" or "stabilisation" approach. In the "splitting" approach first mechanical division of waste takes place and a derived fraction of material is treated biologically. In "stabilisation" approach the entire waste is subject to biological treatment with subsequent splitting of the mass fractions for recycling (metals) and refuse derived fuel (RDF). For the modelling of aerobic MBP within the LCA-IWM tool a technology based on the "splitting" concept was selected. The purpose of splitting is to ensure material and energy recovery and minimization of final disposal.

9.5.1 Process description

The selected technology is fully encapsulated and consists of mechanical pre-treatment with separation of the high caloric light fraction and biological treatment of the remaining waste prior to landfilling. The biological process is conducted in an aerated windrow with a weekly turning of the material. Within this option intensive rotting and aerobic stabilisation takes place in the same windrow. The aeration rate is controlled automatically by the temperature in the windrow. This ensures completion of an intensive rotting in the first three weeks and gradually lower process intensity in the following weeks. Assuming optimal process conditions a far reaching stabilisation of the low caloric fraction is achieved within 14 – 16 weeks. In optimal case the intensive rotting, in which the main part of the decomposition takes place (app. 80% of total decomposition) can be achieved in approx. 4 – 6 weeks (Müller 2001). The end product of these processes - stabilised low calorific fraction - can be landfilled or (if applicable) used for recultivation of degraded land. The high caloric fraction after refining can be either used as a Refuse Derived Fuel in a cement kiln or for energy recovery in an incineration plant. The modelling is based on results of a joint study on the performance of MBP plants in Germany (Soyez et al. 2001a).

Process parameters used for modelling of the MBP are described in the Section 8.8 of Deliverable 3.

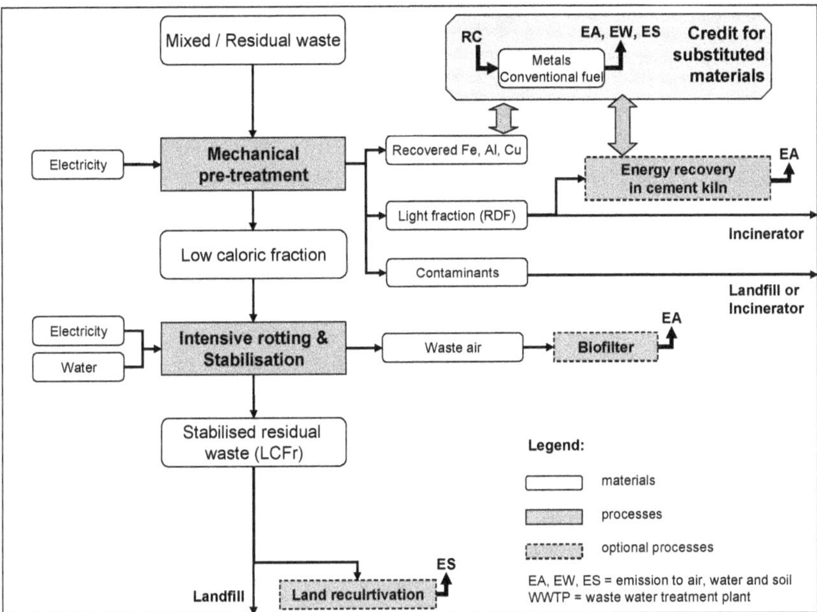

Figure 13. Schedule of the main material flows in the modelled aerobic MBP plant

9.6 Anaerobic Mechanical-Biological Pre-Treatment

In an anaerobic MBP plant residual or mixed waste is treated. Anaerobic MBP is a treatment method which is not widely applied yet, but recently more facilities can be observed (Kern 2001). When compared to aerobic MBP, anaerobic MBP has a number of advantages:

- net energy production
- shorter biological treatment time
- less odorous emissions (reduced need for biological air purification) due to the incineration of biogas.

On the other hand, the anaerobic technology is more complicated and requires higher capital investments (Zeschmar-Lahl and Jager 2000).

9.6.1 Process description

For the modelling of the overall anaerobic MBP the same concept as for the aerobic one was chosen. In Figure 14 a flow schedule of the modelled anaerobic MBP is shown, which is mainly based on the facility in Bassum, Germany. Analogue to the aerobic MBP waste undergoes firstly a mechanical treatment step in which contaminants and a high caloric light fraction are separated. The resulting low caloric fraction is treated biologically. The biological treatment step consists of a fermentation process, in which organic matter is decomposed under anaerobic conditions, producing biogas (methane and carbon dioxide). After approximately three weeks in the one-stage dry thermophile fermenter the fermented waste is stabilised aerobically. Optionally the medium coarse low caloric fraction can be treated fully aerobically in an intensive rotting. The high caloric fraction is used for energy recovery, analogically to the product of aerobic MBP. Parameters applied for modelling of the anaerobic MBP are described in the Section 8.9 of Deliverable 3.

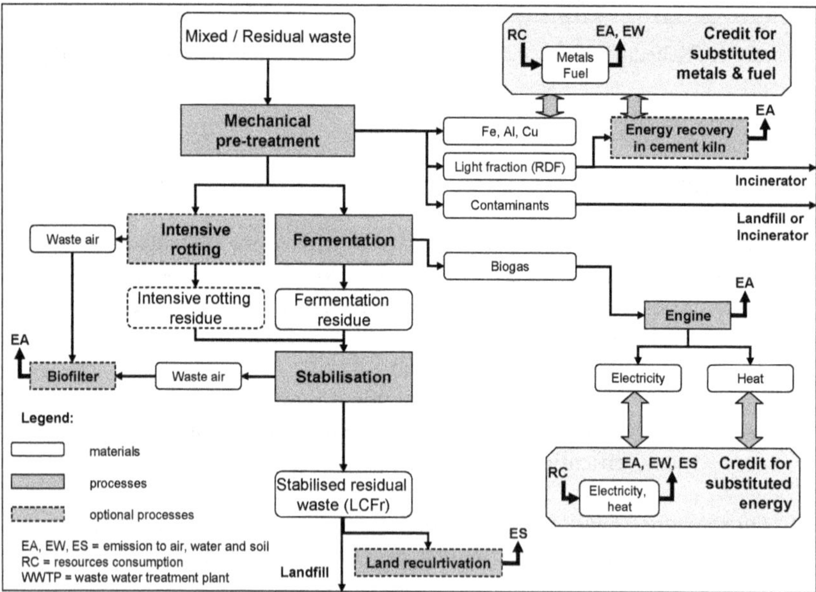

Figure 14. Schedule of the main material flows in the modelled anaerobic MBP plant

9.7 Incineration

Within the incineration process substances contained in waste are oxidised. Burnable waste is in this way transformed into gaseous substances, while inert waste fractions remain as a solid residue in form of incineration slag and ashes. The share of municipal waste which undergoes incineration varies in the old EU Member States from null in Greece to 95% in Luxemburg and almost 100% in Switzerland (Johnke 2003). Waste incineration has a number of environmental benefits: reduction of the waste volume for final disposal, the recovery of energy from waste and reduction of emissions from final waste disposal.

9.7.1 Process description

Incineration technology chosen for modelling is equipped with grate firing. The plant construction (water cooled grate, flue gas recirculation) represents best available technology for municipal waste incineration and allows low surplus air values resulting in low gas volume (ifu and ifeu 2001). The slag can, after pre-treatment, be used as construction material.

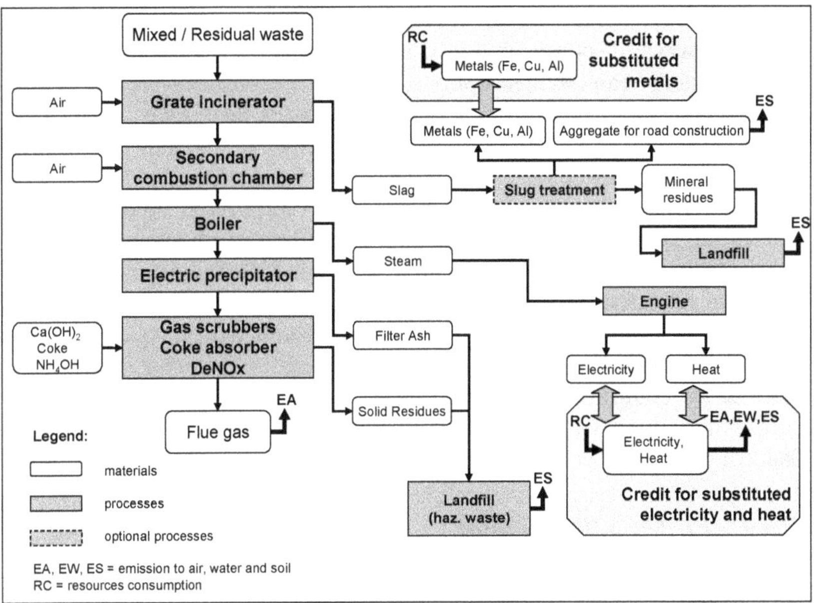

Figure 15. Schedule of the main material flows in the modelled Incineration plant

Energy from waste is recovered in a process and electricity as well as heat are produced. Flue gas cleaning consists of the following steps: (i) an electrostatic precipitator where dust and flue ashes are separated (and disposed as hazardous residues); (ii) an acid flue gas scrubbing facility for removal of hydrogen fluoride, hydrogen chloride and heavy metals; (iii) a neutral sulphur dioxide scrubbing facility with suspended calcium hydroxide; (iv) filter with coke absorber for removal of dioxines/furans and (v) selective catalytic reduction (SCR-catalyser) for denitrification.
Optionally recovery of metals from slags can be modelled in a simple mechanical treatment technology, in which Fe, Cu and Al are recovered (Hellweg 2000). After metals recovery, slags may be used for construction. Alternatively slags can be landfilled. In Figure 15 main material flows of the modelled Incineration plant are presented.

Parameters of the incineration process are described in the Section 8.10 of Deliverable 3.

9.8 Landfilling

Landfilling is an unavoidable element of any waste management system. It is also a waste disposal method which can deal with all materials in the solid waste stream. Other options, such as mechanical-biological pretreatment or incineration produce residual wastes themselves which have to be landfilled. In several European countries (UK, Ireland, Spain) landfilling continues to be the principal waste disposal method (McDougall et al. 2001). In the new EU Member States waste disposal methods other than landfilling hardly exist. The challenge of a modern waste management system is to reduce the volume of waste to be landfilled to a minimum. The principal objective of a modern landfill is the safe long-term disposal of waste, both from a health and from an environmental point of view. As there are emissions from the process (landfill gas and leachate), these also need to be controlled and treated as far as possible. To a limited extent landfilling allows valorisation of waste, in terms of energy recovery from landfill gas.

9.8.1 Process description

The module Landfill describes a modern landfill at which domestic waste and waste similar to domestic is disposed of. Input waste includes raw waste and MBP waste. Figure 16 illustrates main flows of the landfill module. The module describes processes within one landfill cell. There are two main sources of emissions from the landfill: leachate emissions and landfill gas emissions. It is assumed that the landfill is equipped with gas and leachate collection systems. In a default setting gas is collected from the beginning of cell operation until 10 years after cell closure. An advanced user of the module can select no gas collection (e.g. to model existing landfill) and adjust the period of gas collection. The collected gas can either be utilised for energy production or alternatively be burned in a flare. Similarly for leachate collection and treatment from the beginning of cell operation until 50 years after cell closure is specified as default. Both availability of leachate collection and treatment and its duration can be adjusted by an advanced user.

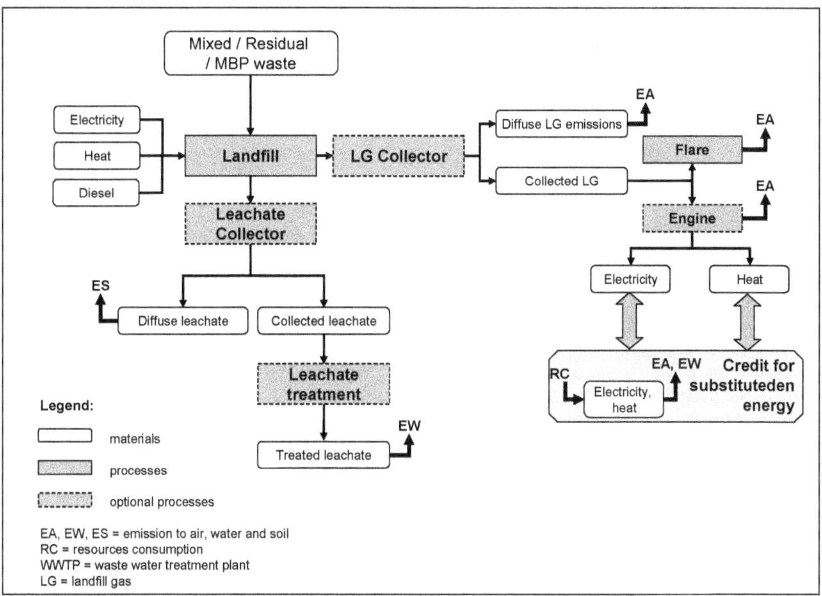

Figure 16. Schedule of the main material flows of the modelled Landfill.

Modelling Parameters of the Landfill module are described in the Section 8.11 of Deliverable 3.

9.9 Waste Recycling

Recovered materials from household wastes which are reprocessed can be used to replace virgin materials, and this may result in overall savings in raw materials, energy consumption and emissions to air, water and soil.

The following modules of the Assessment Tool allow calculating, for each one of the different household waste fractions: paper & cardboard, glass, metals, plastics & composites, packaging-mixed dry recyclables and WEEE, the resulting environmental benefits and/or impacts of some proposed materials recycling processes.

All the following modules consider the environmental loads and impacts of recycling processes and the benefits of avoiding the equivalent or substituting virgin materials production processes.

Modelling parameters of recycling parameters are provided in Chapter 9 of Deliverable 3. Here only a general description of the modelled processes is given.

9.9.1 Paper and cardboard Recycling

This module allows calculating the environmental inventory of recycling paper and cardboard taking into account their substituting production processes from a wood matrix (virgin material). It is assumed that paper and cardboard separately collected are composed by two major sub-fractions: i) paper of de-inking quality and ii) cardboard; and a certain level of contamination. Default composition of separately collected paper and cardboard is provided in Table 20.

Table 20. Composition of separately collected paper & cardboard (default values based on Ludwig C. et al 2003)

Sub-fraction	Composition [%]
Paper of de-inking quality	60
Cardboard	35
Contamination	5

Paper of de-inking quality and cardboard are sorted and pre-cleaned in a Material Recovery Facility (MRF) and transported to the final recycling facilities. Rejects of sorting processes can be landfilled and/or incinerated (user input).

Paper of de-inking quality is assumed to be recycled into newspapers. This process consists of re-pulping and de-inking the incoming paper.

Newspaper production from a wood matrix (virgin material) is assumed to be the avoided process, which includes management of forests, required transport, wood grinding and the complete thermomechanical pulping process.

Cardboard is assumed to be recycled into corrugated board. This process consists of re-pulping the incoming cardboard by a combination of testliner and wellenstoff processes. Corrugated board production from a wood matrix (virgin material), including required transport, is assumed to be the avoided process. This process includes a combination of kraftliner, testliner, wellenstoff and semichemical pulping and re-pulping processes of a wood matrix and recovered paper and cardboard. Data for the production of 100% virgin corrugated board are not usually available, since all corrugated board production processes incorporate some recovered paper and cardboard. It has been possible to estimate the 100% virgin corrugated board scenario by extrapolating linearly from the data for different recycling rates.

9.9.2 Glass Recycling

This module allows calculating the environmental inventory of recycling glass taking into consideration its substituting production process from virgin raw materials. Glass separately collected can be composed by four major sub-fractions: i) mixed glass, ii) green glass, iii) brown glass and iv) clear glass; and a certain level of contamination or impurities. Table 21 shows default composition of separately collected glass.

Table 21. Composition of separately collected glass (default values)

Sub-fraction	Composition [%]
Mixed glass	37
Green glass	20
Brown glass	20
Clear Glass	20
Contamination	3

Glass (all colours) is cleaned and crushed into cullet (broken glass) in a MRF and transported to the final recycling facility. Rejects of cleaning and crushing processes can be landfilled and/or incinerated (user input).

Glass cullet (all colours) is assumed to be recycled again into glass. This process consists of re-melting the incoming cullet in a furnace. Melting process of raw materials, including required transport, is assumed to be the avoided process. Data for production of 100% virgin glass are not usually available, since all glass production processes incorporate some recovered glass. It has been possible to estimate the 100% recycled glass scenario by extrapolating linearly from the data for different recycling rates.

9.9.3 Metals Recycling

This module allows calculating the environmental inventory of recycling metals taking into consideration their substituting production process from virgin raw materials. It is assumed that metals separately collected are composed by two major sub-fractions: i) tinplate steel and ii) aluminium; and a certain level of contamination or impurities (for default composition see Table 22).

Table 22. Composition of separately collected metals (default values based on RCD-Env. & Pira Int. 2003)

Sub-fraction	Composition [%]
Tinplate steel	77,5
Aluminium	17,5
Contamination	5

Metals are sorted in a MRF and transported to the final recycling facilities. Rejects of sorting processes can be landfilled and/or incinerated (user input).

Tinplate steel is assumed to be recycled into secondary steel by an electrical re-melting process and further metallurgical processes. Primary steel production from raw materials, including required transport, is assumed to be the avoided process, which includes a blast furnace melting process and further metallurgical processes.

Aluminium is assumed to be recycled into secondary aluminium. This process consists of grinding, de-coating and re-melting the incoming aluminium. Primary aluminium production from raw materials, including required transport, by an electrolysis process is assumed to be the avoided process.

9.9.4 Plastic and composites Recycling

This module allows calculating the environmental inventory of recycling plastics and composites taking into account their substituting production processes from raw materials. It is assumed that plastic and composites separately collected are composed by six major sub-fractions: i) high density polyethylene (HDPE) ii) polyethylene terephthalate (PET), iii) low density polyethylene (LDPE), iv) mixed plastics, v) liquid beverage cartons (LBC) and vi) other composites; and a certain level of contamination or impurities. Default composition of separately collected plastics and composites in presented in Table 23.

Table 23. Composition of separately collected plastics and composites (default values based on RCD-Env. & Pira Int. 2003)

Sub-fraction	Composition [%]
HDPE (all colours)	13,1
PET	15,9
LDPE film	16,9
Mixed plastics	25,3
Liquid Beverage Cartons (LBC)	11,2
Other composites	6,6
Contamination	11

Plastics and composites are sorted in a MRF and transported to the final recycling facilities. Rejects of sorting processes and the sub-fraction "other composites" can be landfilled and/or incinerated (user input).

HDPE (all colours) is assumed to be recycled into HDPE multi-layered bottles. This process consists of grinding, hot cleaning and granulating the incoming HDPE and its co-extrusion with virgin HDPE. Primary HDPE production from virgin raw materials, including required transport, obtained by a simple extrusion process is assumed to be the avoided process.

PET is assumed to be recycled into PET three-layered bottles. This process consists of regenerating by heating the incoming PET and its injection with virgin PET (three-layer). Primary PET production from virgin raw materials, including required transport, obtained by a mono-layer injection process is assumed to be the avoided process.

LDPE film is assumed to be recycled into LDPE sacs. This process consists of cleaning and granulating the incoming film and its co-extrusion with virgin LDPE. Primary LDPE production from virgin raw materials, including required transport, obtained by a simple extrusion process is assumed to be the avoided process.

Mixed plastics are assumed to be recycled into plastic pickets. This process consists of heating and adding calcium carbonate to the incoming mixed plastics and extruding them. Wood pickets production is assumed to be the avoided process, which includes forests management, required transport, pickets production and treatment with Cu, Cr and As.

Liquid beverage cartons (LBC) are assumed to be recycled into pulp for domestic paper. This process consists of re-pulping and de-inking the incoming liquid beverage cartons (recycling of aluminium and polyethylene are not considered). Production of primary pulp is assumed to be the avoided process, which includes forest management, required transport and chemical pulping process with sodium sulphide (Kraft process) non bleached.

9.9.5 Packaging Recycling - Mixed dry recyclables

The Assessment Tool also allows the possibility for calculating the environmental inventory of recycling packaging-mixed dry recyclables minus their substituting production processes from virgin raw materials. This module is a composition and an alternative to the four previous ones: paper and cardboard, glass, metals and plastics and composites (see Sections 9.9.1- 9.9.4).

Consequently, packaging-mixed dry recyclables separately collected can be composed by fourteen major sub-fractions: i) paper of de-inking quality, ii) cardboard, iii) mixed glass, iv) green glass, v) brown glass, vi) clear glass, vii) tinplate steel, viii) aluminium, ix) high density polyethylene (HDPE), x) polyethylene terephthalate (PET), xi) low density polyethylene (LDPE), xii) mixed plastics, xii) liquid beverage cartons (LBC) and xiv) other composites; and a certain level of contamination or impurities.

Packaging-mixed dry recyclables are sorted and pre-cleaned in a MRF and transported to the final corresponding recycling facilities (see Sections 9.9.1- 9.9.4). Rejects of sorting processes can be landfilled and/or incinerated (user input).

9.9.6 Waste electrical and electronic equipment

Detailed modelling of recycling and disposal of Waste electrical and electronic equipment (WEEE) is very complex and is far out of the scope of this tool and project. Thus within this module a simplified approach is taken, based on literature data of the LCA study for WEEE performed in the United Kingdom in 1999 (Ecobalance UK and DMG Consulting Ltd. 1999). The output of this study covers only a part of the environmental aspects covered within the other modules. Nevertheless, it is worth considering, since it provides a good insight into environmental benefits of recycling of WEEE as opposed to landfilling.

This module allows calculating the environmental inventory of recycling of WEEE taking into consideration their substituting production processes from virgin raw materials. The landfilling of not separately collected WEEE is considered in the Landfill module.

It is assumed that separately collected WEEE is only composed of eight major and representative appliances: i) refrigerator (including fridge-freezer), ii) personal computer (monitor, processor and keyboard), iii) washing machine, iv) vacuum cleaner (including held appliances), v) telephone, vi) television (including large and small screens), vii) lawnmower and viii) kettle.

This WEEE composition contains a list of electrical and electronic equipment which together comprise almost 50% of the WEEE generated in the United Kingdom in 1998. Additionally, these products represent a broad range of potential problems due to hazardous substances content. This default composition can be modified by the user of the tool.

Table 26. Representative composition of separately collected WEEE (default values based on Ecobalance UK and DMG Consulting Ltd. 1999)

Sub-fraction	Composition [%]
Refrigerator (including fridge-freezer)	21
PC (monitor, processor and keyboard)	26
Washing machine	34
Vacuum cleaner (including held appliances)	3
Telephone	1
Television (large and small screen)	9
Lawnmower	5
Kettle	1

For calculating the corresponding environmental inventories of recycling or landfilling each of the different mentioned product categories, only impact data for climate change (kg CO_2 equiv.), acidification (kg SO_2 equiv.) and eutrophication (kg PO_4 equiv.) have been considered. For the abiotic depletion recovery of metals from WEEE has been accounted for. Recycling of other parts of the equipment has not been taken into consideration.

Recycling processes of WEEE include pre-treatment, transport of materials to their corresponding reprocessing facilities, recycling processes and appropriate treatment-disposal of rejects and/or hazardous components-substances and the corresponding avoided processes for producing primary materials from virgin raw materials.

9.10 Production of energy and auxiliary materials

9.10.1 *Electricity*

The electricity inventories of all previously considered processes (waste treatment, disposal, recycling & primary material manufacturing) are transformed into environmental loads by considering the selected and country-specific electricity supply mix (Ecoinvent-2000 2003). In case the considered scenario results in a net electricity consumption for the total waste management system, the environmental burdens of electricity production of primary sources are considered. In case of a net electricity production in the total waste management system, this secondary electricity is assumed to substitute electricity of primary sources. Thus, the avoided environmental burdens (of electricity production from primary sources) are accounted for credits.

Life cycle inventories (LCI) of country-specific electricity supply mixes are available in the tool for the following European countries: AT (Austria), BE (Belgium), CH (Switzerland), ES (Spain), YU (Yugoslavia), FR (France), GR (Greece), IT (Italy), LU (Luxemburg), NL (The Netherlands), PT (Portugal), DN (Denmark), DE (Germany), FI (Finland), GB (United Kingdom), IE (Ireland), SE (Sweden), NO (Norway), CZ (Czech Republic), HU (Hungary), PL (Poland), SK (Slovak Republic), SI (Slovenia), HR (Croatia), BA (Bosnia-Herzegovina) and MK (Macedonia).

These LCIs include the shares of domestic electricity production by technology and imports from neighbouring countries (production mixes) at the busbar. These inventories do not include transformation, transport or distribution losses. Net domestic electricity production and import shares are based on annual averages.

9.10.2 *Heat*

As in the case of electricity, the considered scenario can either result in a net heat consumption or production. The environmental burdens caused by the supply of heat from primary sources are either added to or subtracted (credits) from the total environmental burdens of the scenario. The user will be able to choose the relative share of the fossil fuels producing the heat. Average values for the supply of heat from hard coal, natural gas, heavy fuel oil and light fuel oil are available.

9.10.3 Auxiliary materials

Steel and lubricants

In some of the modules the consumption of lubricants or steel parts, due to maintenance of the facilities, occurs. Based on the total amounts of lubricant and steel parts consumed for the considered scenario and the specific contribution per kg or litre the total environmental impact of auxiliary materials is calculated.

Diesel for transport and for machinery

In many of the modules the consumption of diesel occurs. Diesel use has been divided in diesel for transportation and diesel for machinery. It is assumed that the environmental burdens per litre of diesel are equal for all machinery processes and transportation processes respectively. The total amount of both kinds of diesel consumption in the scenario is taken to determine the environmental burdens connected with both the production (identical for both kinds of diesel) and combustion of diesel.

10 Case Studies

10.1 Optimal waste management scenarios for the selected cities from fast growing European regions –introduction

Main Objectives:

- to find the best waste management scenarios among analysed for selected European cities from regions with rapid growing economies and to verify the developed tools in practice by the Partners from the respective cities,
- to apply the developed tools for waste management planning by the potential users and to verify their practicability,
- to test wide range of alternative solutions in order to show possibilities and advantages of the tools developed.

Four to five alternative waste management scenarios were developed for each of five European cities (see Figure 17) in close collaboration with the local municipality representatives.

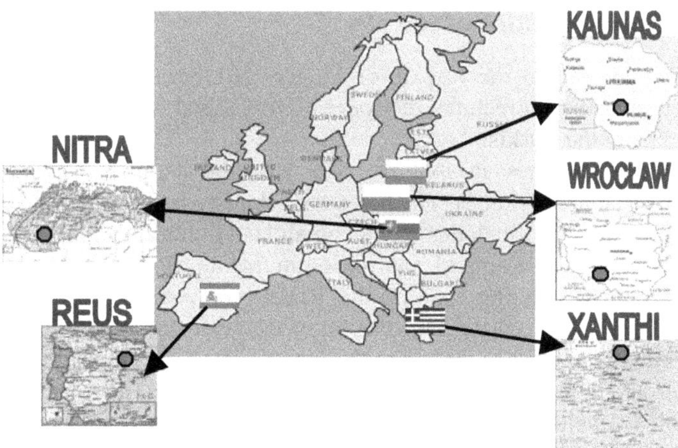

Figure 17. Locations of cities

Those scenarios reflect various situations and predictions of the waste management in the target cities. One scenario (generally No. 1) characterises the present situation, while the other three or four analyse some possible developments (characterised by different solutions of Temporary Storage, Collection, Transport and Treatment). Obviously, there are also many other potential alternatives in each case, not presented here due to the limited volume of the handbook, which should be assessed using developed tools in order to find the best solution. Presented results of case studies show also the wide range of different potential solutions and demonstrate possibilities of the tool to assess them. Among numerous tables and graphs provided by the Tool for each scenario assessed only a few have been selected (for each municipality) for presentation in this chapter. They summarise and generalise the results of the studies conducted and show the most important data and relationships.

10.2 Selected cities with rapid growing economies

10.2.1 Xanthi-Greece

Short characteristics of the city

Xanthi is located in the Region of Thrace (Northern Greece), bordering Bolgaria and the Prefectures of Rodopi, Drama and Kavala, and is about 100 km from the Turkish border.

The Prefecture has an area of 1.793 km^2 and its population is about 102.000 inhabitants (2001). The population density is about 56 inh/km^2.

Xanthi Prefecture includes 6 Municipalities (Xanthi, Avdira, Vistonida, Myki, Stavroulopi, Topiros) and 4 Communes (Thermes, Kotyli, Satres, Selero). Situated at the foot of the Rodopi mountains, city of Xanthi (45.200 inhabitants, 2001) is the capital of the Prefecture.

After a tormented past linked to the history of Thrace, the city grew as a commercial, cultural and spiritual centre in the region. The tobacco Industry, which developed there from the 18th century, helped to spread its renown across Europe and led towards its prosperity. Democritus University is a real growth engine for regional economic activity.

Characteristics of municipal solid waste quantities and composition

The characteristics of the MSW generated in Xanthi over the next 10 years, as projected by the Prognostic Tool, are shown in Table 24 and Table 25 The base year is 2003. The quantities are presented in Table 24 while Table 25 shows the composition of MSW (in % by weight) by main fraction, namely paper and cardboard, glass, organic waste, metals, plastics and composities, etc.

Table 24. Future generated MSW quantities in the city of Xanthi

Year	2003	2004	2005	2006	2007	2008	2009
Quantity (kg/cap./year)	304	312	319	325	331	338	345
Year	2010	2011	2012	2013	2014	2015	
Quantity (kg/cap./year)	352	359	366	373	381	389	

Table 25. MSW composition in the city of Xanthi (projections)

	Components of MSW (% by weight)								
Year	Organic	Paper & cardboard	Glass	Metals	Plastics & composites	Hazardous waste	WEEE	Other	Bulky waste
2003	45	21	5	5	9	3	1	10	1
2008	44	22	5	5	10	3	1	10	1
2015	42	23	5	5	10	3	1	10	1

Existing municipal waste management situation

The Solid Waste Management Authority of Xanthi Prefecture is the authority responsible for the management of all types of solid waste in the Prefecture of Xanthi. In 2003, there were about 40.000 tons (the collected quantity of solid waste of the city of Xanthi was 13.700 tons) of commercial and household solid waste per year, in addition to industrial, agricultural, and other solid waste. Regarding domestic waste, the Solid Waste Management Authority is responsible for collection and landfilling. All domestic waste is landfilled on a landfill situated 10 km from the city of Xanthi.

Scenario 1:
Temporary Storage - Collection & Transport: only one waste stream (residual waste), no source separation.
Treatment - Disposal: landfilling of residual waste.

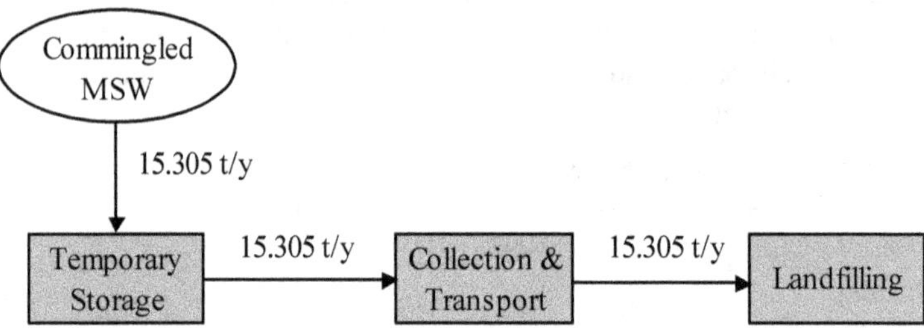

Figure 18. Current waste management system in Xanthi (Scenario 1)

Description of chosen waste management scenarios for Xanthi.

Scenario 2:
Temporary Storage - Collection & Transport: only one waste stream (residual waste), no source separation.

Treatment - Disposal: incineration of residual waste, landfilling of incineration hazardous residues.

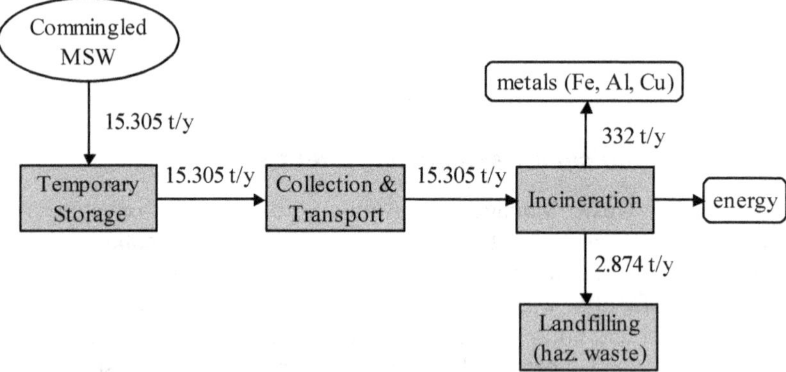

Figure 19. Alternative MSWMS for Xanthi - Scenario 2

Scenario 3:

Temporary Storage - Collection & Transport: separate collection of bio-waste and residual waste.

Treatment - Disposal: anaerobic digestion of bio-waste and landfilling of residual waste.

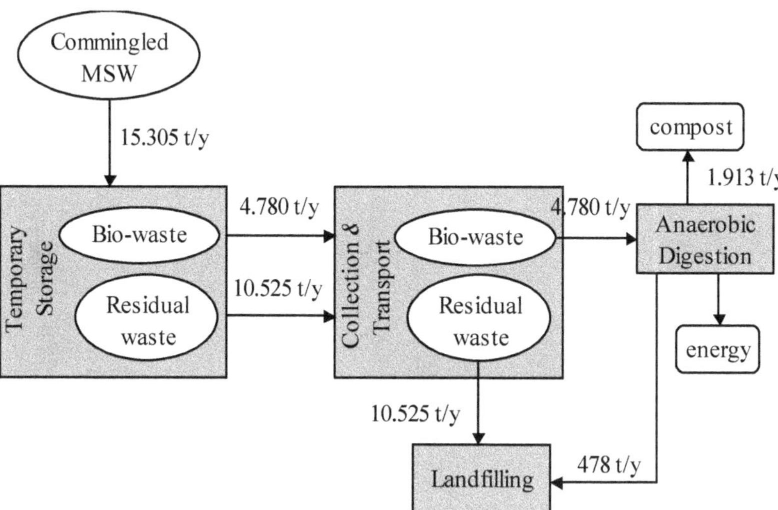

Figure 20. Alternative MSWMS for Xanthi - Scenario 3

Scenario 4:

Temporary Storage - Collection & Transport: separate collection of mixed dry recyclables (MDR: paper & cardboard, glass, metals, plastics & composites) and residual waste.

Treatment - Disposal: sorting of mixed dry recyclables and landfilling of residual waste.

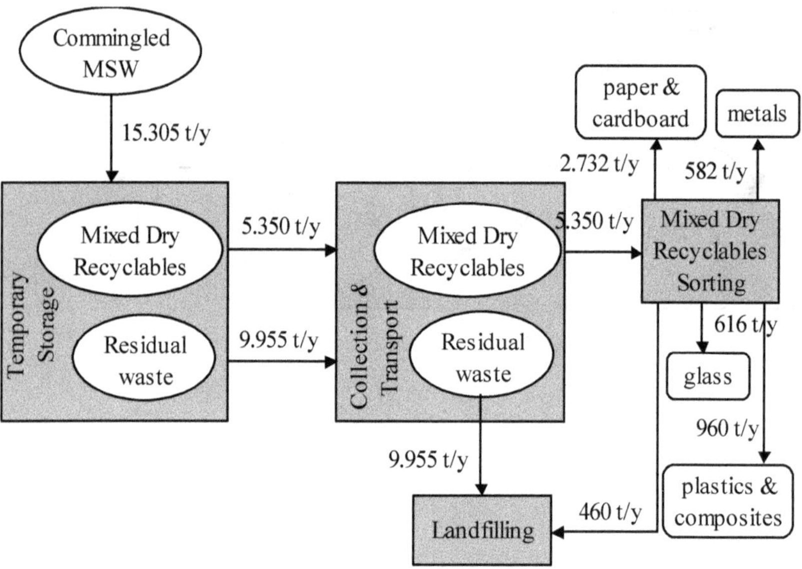

Figure 21 Alternative MSWMS for Xanthi - Scenario 4

Scenario 5:

Temporary Storage - Collection & Transport: separate collection of bio-waste, mixed dry recyclables and residual waste.

Treatment - Disposal: anaerobic digestion of bio-waste, sorting of mixed dry recyclables and landfilling of residual waste.

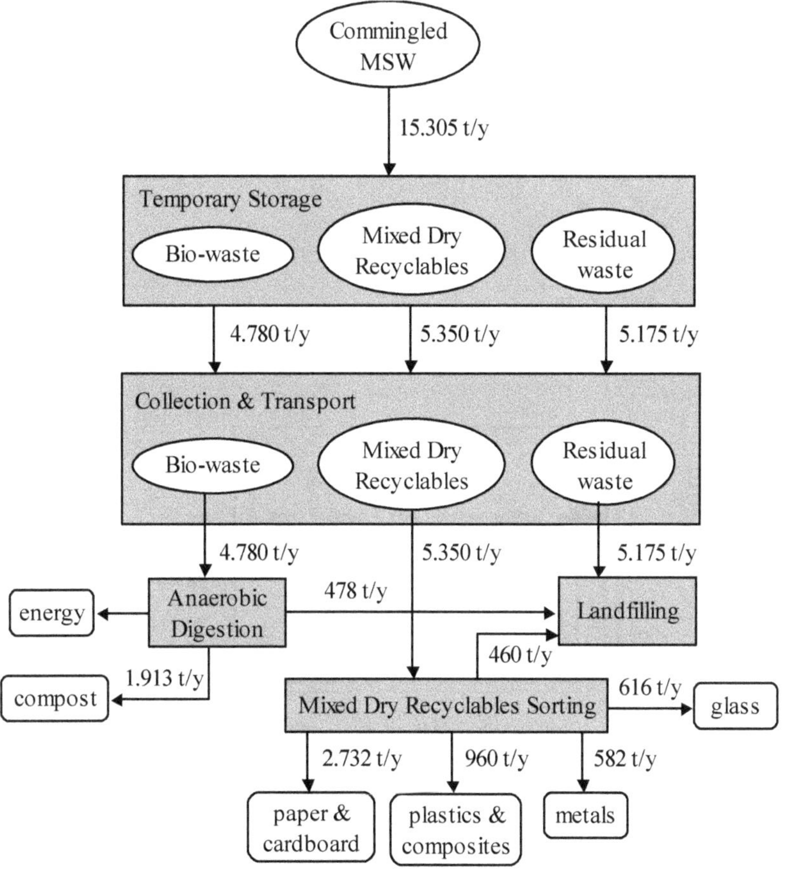

Figure 22 Alternative MSWMS for Xanthi - Scenario 5

Outputs from the usage of tools for scenarios - results of modeling

Environmental Assessment

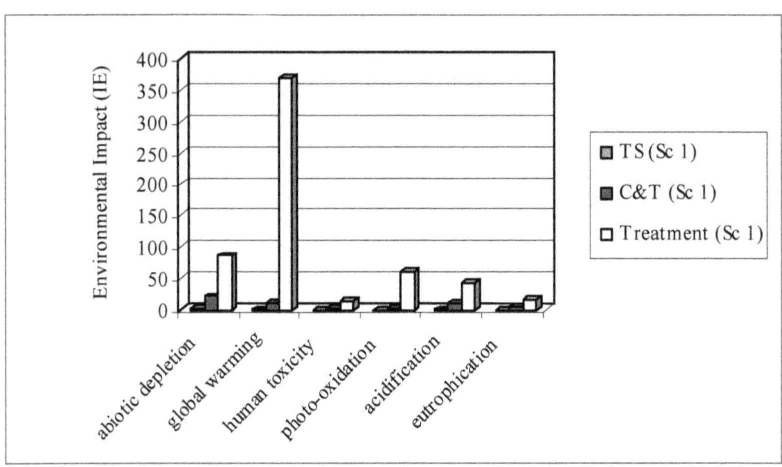

Figure 23. Environmental impacts of each subsystem (Scenario 1 - Xanthi)

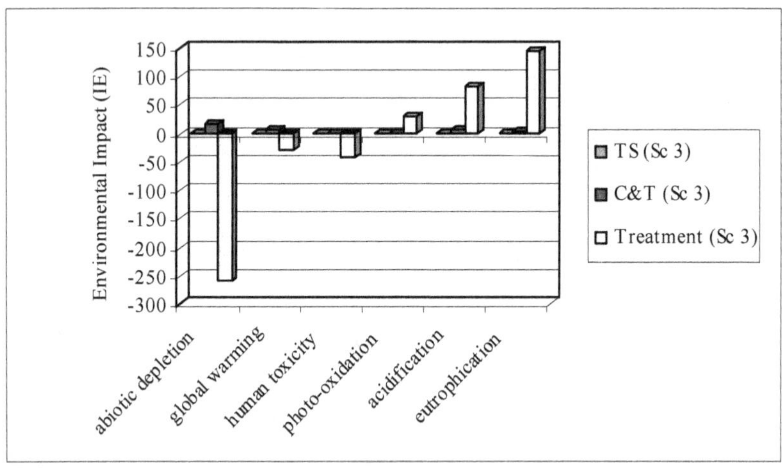

Figure 24. Environmental impacts of each subsystem (Scenario 2 - Xanthi)

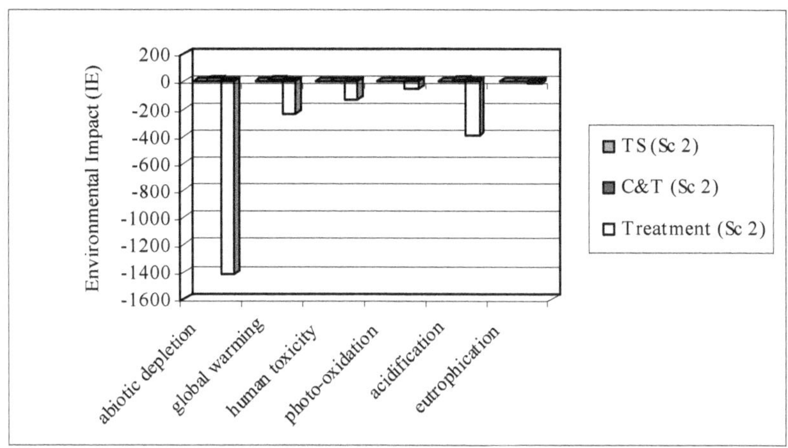

Figure 25. Environmental impacts of each subsystem (Scenario 3 - Xanthi)

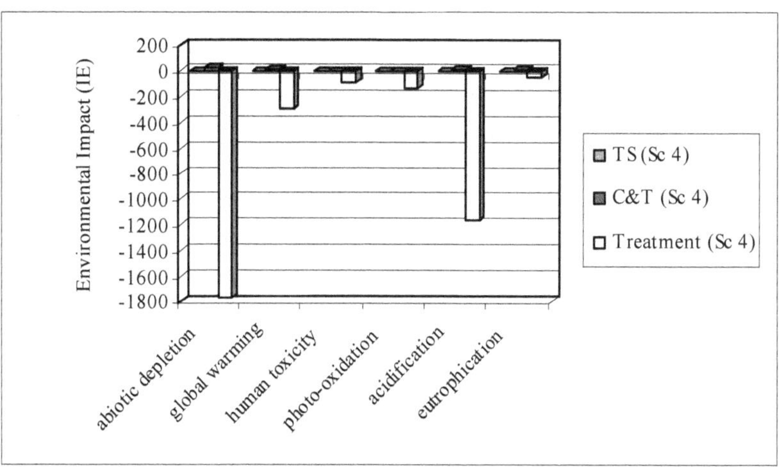

Figure 26. Environmental impacts of each subsystem (Scenario 4 - Xanthi)

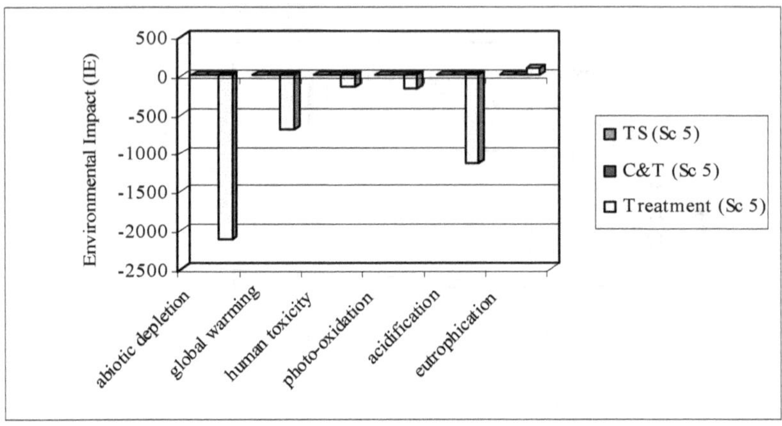

Figure 27. Environmental impacts of each subsystem (Scenario 5 - Xanthi)

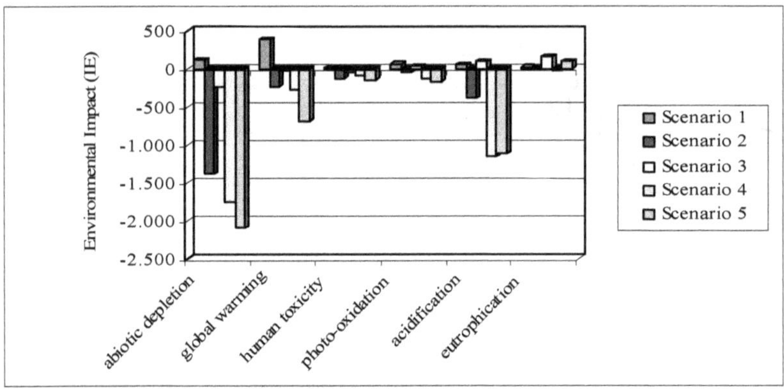

Figure 28. Environmental impacts of MSWMS scenarios for Xanthi

Table 26. Environmental Impacts of MSWMS in Xanthi

Subsystem	Scenarios				
	Sc 1	Sc 2	Sc 3	Sc 4	Sc 5
			Abiotic depletion (IE)		
Temp. Storage	0,79	0,79	0,97	3,39	3,59
Coll. & Transport	20,96	20,96	17,17	19,93	15,58
Treatment	86,13	-1.411,67	-257,41	-1768,80	-2.113,18
			Global warming (IE)		
Temp. Storage	0,25	0,25	0,25	0,66	0,66
Coll. & Transport	9,64	9,64	7,90	9,16	7,16
Treatment	368,20	-240,88	-27,55	-291,76	-693,83
			Human toxicity (IE)		
Temp. Storage	0,01	0,01	0,01	0,01	0,01
Coll. & Transport	0,74	0,74	0,61	0,70	0,55
Treatment	12,40	-125,55	-43,51	-90,17	-146,06
			Photo-oxidation (IE)		
Temp. Storage	0,03	0,03	0,03	0,07	0,07
Coll. & Transport	1,15	1,15	0,94	1,09	0,85
Treatment	61,26	-57,73	31,54	-140,39	-171,14
			Acidification (IE)		
Temp. Storage	0,16	0,16	0,18	0,60	0,62
Coll. & Transport	9,79	9,79	8,02	9,31	7,28
Treatment	41,94	-401,74	82,42	-1.160,97	-1120,94
			Eutrophication (IE)		
Temp. Storage	0,03	0,03	0,04	0,14	0,14
Coll. & Transport	4,58	4,58	3,75	4,35	3,40
Treatment	16,76	-7,84	144,08	-42,90	84,36

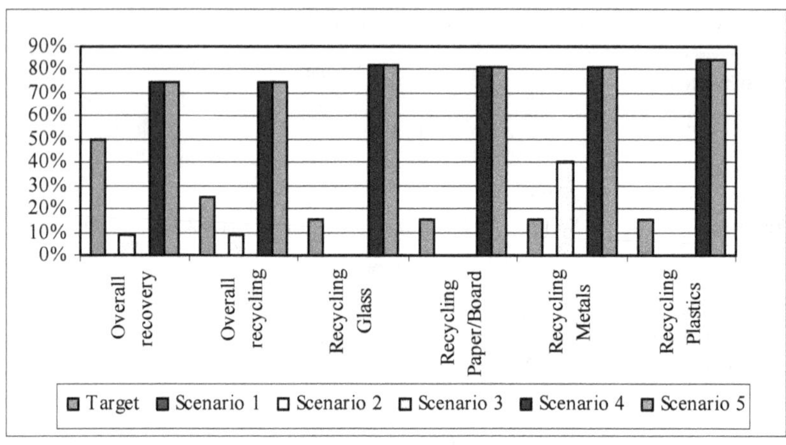

Figure 29. Recycling and recovery of Packaging Waste (Xanthi)

Economic Assessment

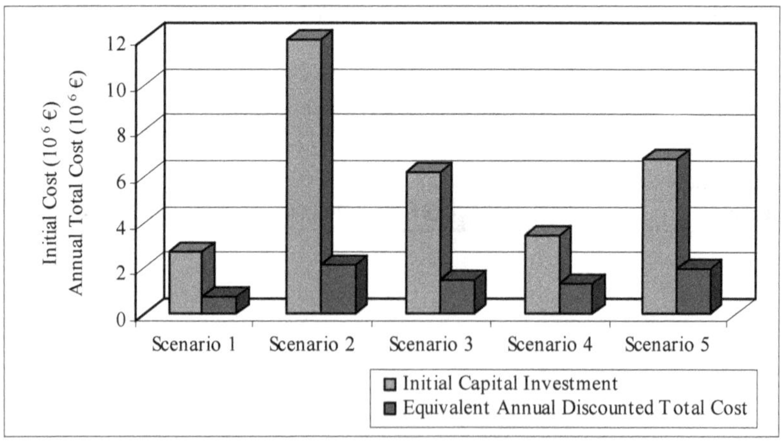

Figure 30. Economic impacts of MSWMS scenarios for Xanthi

Table 27. Economic impacts of MSWMS in Xanthi

	Scenarios				
	Sc 1	Sc 2	Sc 3	Sc 4	Sc 5
Initial Capital Investment (10^6 €)	2,70	12,00[a]	6,20	3,40	6,80
Temp. Storage	3%	1%	2%	7%	4%
Coll. & Transport	19%	4%	8%	19%	8%
Treatment	78%	95%	90%	74%	88%
Equivalent Annual Discounted Total Cost (10^6 €)	0,80	2,15[a]	1,50	1,40	2,00
Temp. Storage	2%	1%	1%	2%	2%
Coll. & Transport	66%	25%	34%	48%	29%
Treatment	32%	74%	65%	50%	69%

[a] Cost for disposal of hazardous residues of incineration is not included.

Table 28. Economic Efficiency at the Municipal level (Xanthi)

Indicators	Scenarios				
	Sc 1	Sc 2	Sc 3	Sc 4	Sc 5
Cost per ton (€/ton)	52,23	140,00[a]	100,00	90,00	130,00
Cost per household (€/hh)	51,00	136,00[a]	97,00	87,00	128,00
Cost per person (€/person)	18,00	47,00[a]	33,00	30,00	45,00
Revenue from recovered material and energy (€)	0	860.000	200.000	390.00	595.000
Total cost as % of GNP of the city (%)	0,40%	1,00%[a]	0,80%	0,70%	1,00%
Diversion between revenue and expenditures for MSWM (%)	80,00%	30,00%[a]	42,00%	47,00%	32,00%

[a] Cost for disposal of hazardous residues of incineration is not included.

Table 29. Economic Efficiency at the subsystem level (Xanthi)

Subsystem	Cost per ton of waste (€/ton)				
	Sc 1	Sc 2	Sc 3	Sc 4	Sc 5
Temporary Storage	0,95	0,95	0,90	2,10	2,00
Collection & Transport	34,60	34,60	33,00	42,00	38,00
Anaerobic digestion	-	-	164,00	-	164,00
MDR sorting	-	-	-	88,20	88,20
Incineration	-	103,25	-	-	-
Landfilling	16,80	-[a]	18,10	18,60	22,60

[a] Cost for disposal of hazardous residues of incineration is not included.

Table 30. Equity and Dependence on Subsidies (Xanthi)

Indicators	Scenarios				
	Sc 1	Sc 2	Sc 3	Sc 4	Sc 5
Cost per person as % of minimum wage	60,00%	157,00%	110,00%	100,00%	150,00%
Cost per person / income per person	0,08%	0,22%	0,15%	0,14%	0,20%
Subsidies per person (€/person)	3,20	12,70	6,80	4,10	7,50

Social Assessment

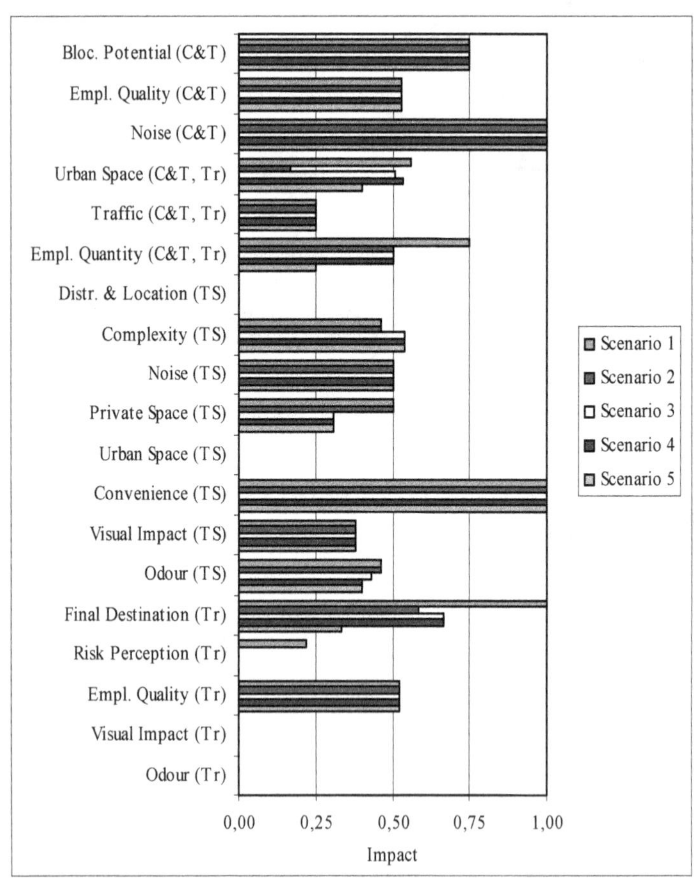

Figure 31. Social Impacts of MSWMS scenarios for Xanthi

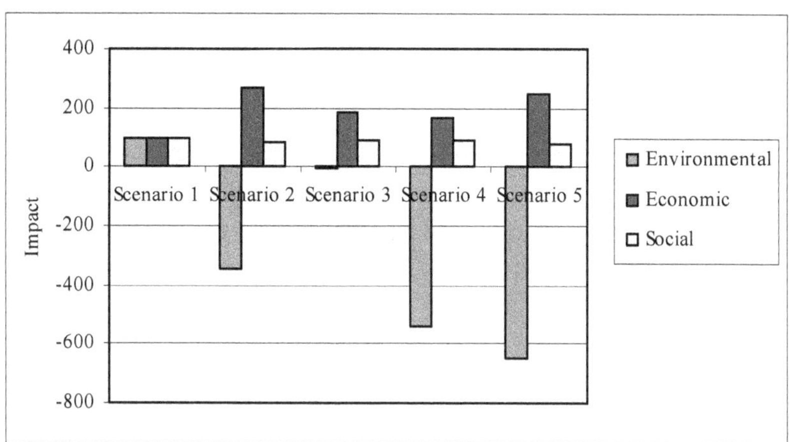

Figure 32. Relative Impacts on sustainability of MSWMS scenarios for Xanthi

10.2.2 Kaunas-Lithuania

Short characteristics of the city

Lithuania - as one of the three Baltic States neighbouring Latvia and Estonia - is situated in the north-east of Europe. Kaunas is the second largest city of Lithuania. Total area of Kaunas is ca. 15,7 thousand ha. According to the population census early in the year 2003, 374.000 inhabitants lived in Kaunas. The average air temperature determined over many years in Kaunas is 6,9° C. The difference between the warmest month July and the coldest month January is around 25° C. The average annual precipitation in Kaunas is 572 mm.

Kaunas is located almost in the center of the country and has good geographical situation with respect to transport network. Highway Via-Baltica connects Warszawa with Helsinki and is presently under modernization. The planned new railway Rail-Baltica will also connect Warszawa and Helsinki. The navigable river Nemunas connects Kaunas with Baltic Sea.

Characteristics of municipal solid waste quantities and composition

Actual and forecasted municipal solid waste generation per capita for Kaunas is shown in Figure 33, while waste composition in 1995-2004 is shown in Figure 34. The previous statistic data shows that the average composition of mixed domestic waste is fluctuating, but the common tendency is a slow increase in the percentage of biowaste, plastics and glass. In Table 31 balance scheme of domestic and domestic similar waste as mass percentage is presented.

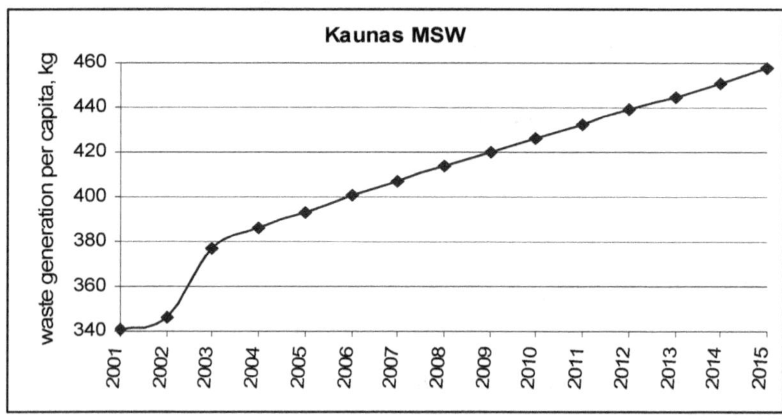

Figure 33. Municipal Solid Waste amounts, actual data and forecast for Kaunas

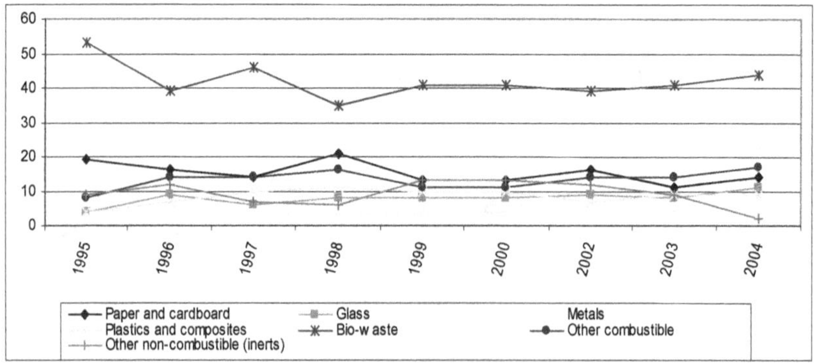

Figure 34. Domestic waste compositions in 1995-2004, mass % (Kaunas)

Table 31. Composition of municipal waste as mass percentage for Kaunas

Year	Organic	Paper & cardboard	Plastics , composites	Glass	WEEE	Metals	Hazardous	Other	Bulky waste
	%								
2003	35,7	17,1	6,8	12,6	2,0	3,2	1,0	21,6	-
2013	33,6	18,9	7,5	12,6	2,0	2,9	1,0	21,5	-

Existing municipal waste management situation

Initiated by the EU, Lithuania began to introduce new laws in accordance with EU requirements: thus, among others, laws on waste management came into force. Table 32 shows environmental laws in Lithuania and their compatibility with EU legislation.

Table 32. Lithuanian waste legislation

Title of environmental law	Compatibility with EU law (Yes/No)
Law on Waste Management	Y
Law on the Management of Packaging and Packaging Waste	N
Public strategic plan for waste management	N
Public program for hazardous waste management	Y
Instruction on landfill construction, exploitation, closure and after closure supervision	Y
National strategic waste management schedule	N
Law on Animal Waste Management	Y
Pesticide waste management rule	N
Law on the Management of Radioactive Waste	Y
Oil waste management instruction	N
Instructions for Waste Management	Y
Medical Waste Management. Lithuanian Hygiene Norm HN 66:2000	Y
Environmental requirements for waste incineration	Y
PCB/PCT waste management instruction	Y
Waste classification	N

The biggest problem in fulfilling EU laws is the lack of specialists and experience in waste management in Lithuania.

Figure 35 shows waste management scheme for Kaunas city. The main waste disposal method is still landfilling. However the hazardous waste and a small part of recyclable materials are transferred for treatment. The biggest problems are waste separation and recycling. It is necessary to improve both the waste management system and the motivation of people to promote waste separation.

Municipality council states the tariffs for waste treatment. The persons living in blockhouses pay 2 Lt/person (about 0,58 Euro) for waste treatment. The private houses holders pay 3,5 Lt/month (about 1 Euro) per 110l container. Wastes are collected 2 times per week in blockhouses, and every 2 weeks in private housing areas.

There are about 21.000 containers used for mixed waste temporary storage in Kaunas city. The 120 - 240l HDPE containers are in use in private housing areas (15%) and 600 – 10000l steel plate containers in blockhouses area.

There are about 1.700 steel plate containers 1200 – 10000l used for separately collected waste in blockhouses area. Among these about 57% are in use for glass collection.

7- and 3-tons capacity vehicles are used for waste collection and 26- and 14- tons capacity vehicles are used for waste transportation from transfer station to the landfill.

There is transfer station operated by Kaunas city. Wastes from collection vehicles are unloaded, pressed and reloaded to vehicles with bigger capacity to be transported to landfill. The separately collected waste is there sorted and sent to recycling to others companies.

In composting plant only green waste from parks and squares is composted; there is no composting of household biowaste.

The landfill of Kaunas Municipality is situated near the small town Lapes, 10 km northeast from Kaunas. Landfills area is 374.000m^2. Landfill was constructed in 1973. Landfill fulfills European legislation standards and is equipped with a leachate treatment facility. About 139.000 t waste per year will be landfilled untill 2014. Landfill serves not only Kaunas city but other cities of Kaunas region also.

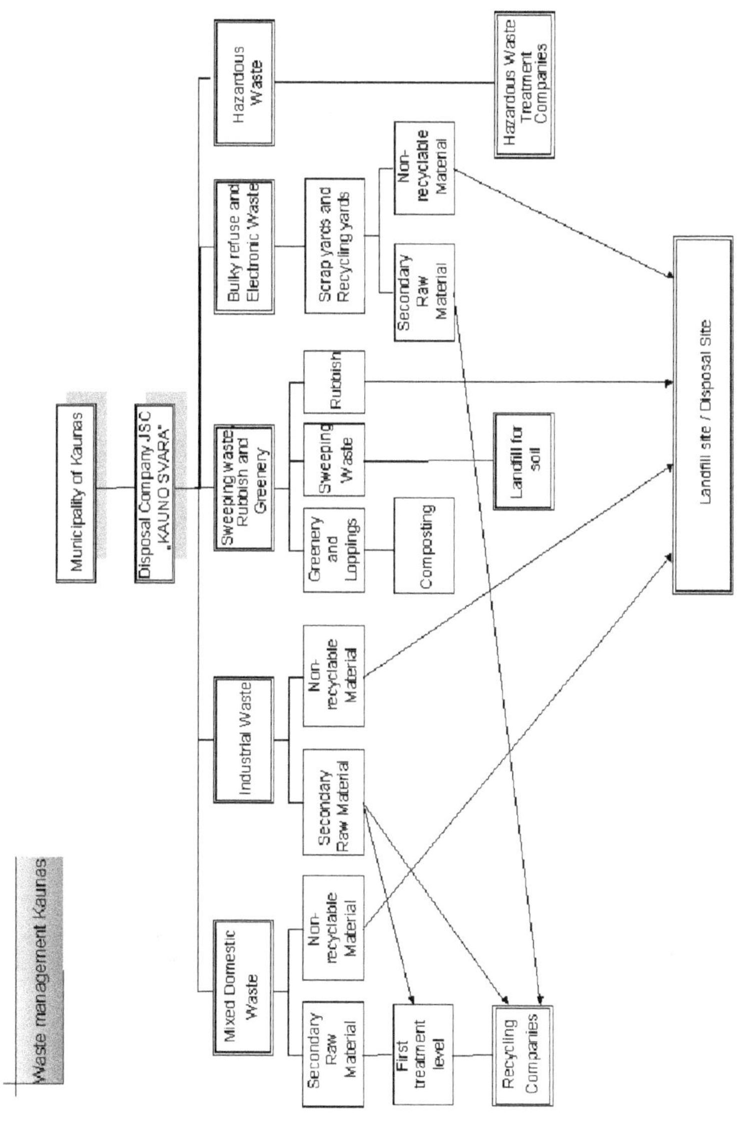

Figure 35.Current waste management scheme of Kaunas Municipality

Description of chosen waste management scenarios for Kaunas.

In 2004 studies on waste management plans for Kaunas region were completed. They showed the necessity to reduce waste generation. However this study did not clearly define which treatment way should be chosen (composting or incineration). In this study, mechanical-biological pre-treatment and incineration as alternative treatment options have been chosen to see how much they would contribute to economic, environmental and social sustainability of MSWMS. The chosen waste management scenarios can be understood as stepped evolution of waste management system, i.e. the next scenarios are an extension of the previous scenarios. It was decided that collection and transport subsystems will be the same for each scenario. Only the numbers of containers will be increased to achieve higher rates of separate waste collection.

Scenario 1: Separate collection and sorting of glass, plastic, paper & cardboard and metals at a low rate, landfilling of all other waste streams after reloading at the transfer station (current situation);

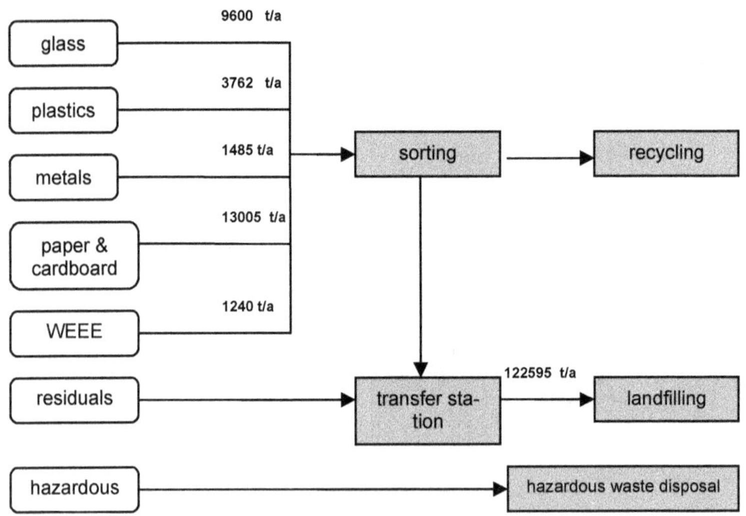

Figure 36.Scenario 1 - Kaunas

Scenario 2: Increased rate of separate collection and sorting of glass, plastic, paper & cardboard and metals, landfilling of all other waste streams after reloading at the transfer station.

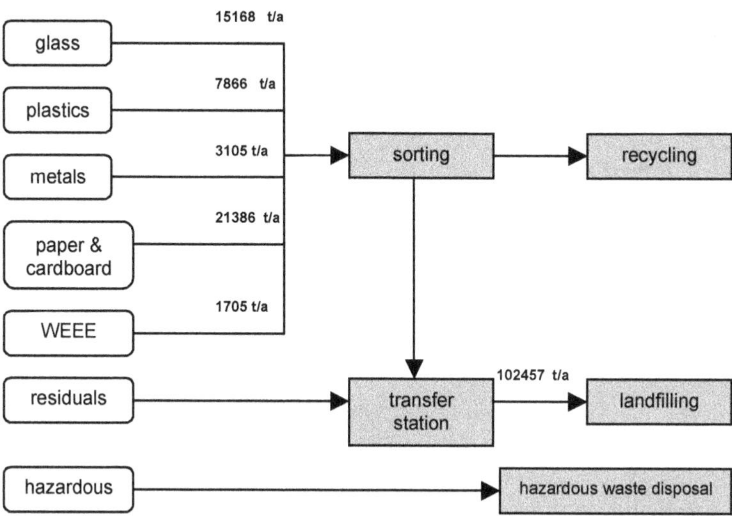

Figure 37. Scenario 2 – Kaunas

Scenario 3: Increased rate of separate collection and sorting of glass, plastic, paper & cardboard and metals, aerobic mechanical – biological pre-treatment of the residual waste before landfilling.

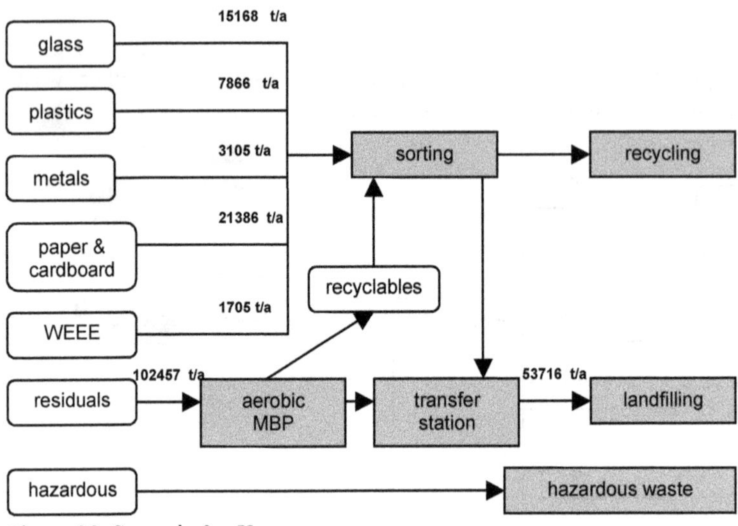

Figure 38. Scenario 3 – Kaunas

Scenario 4: Increased rate of separate collection and sorting of glass, plastic, paper & cardboard and metals, aerobic mechanical – biological pre-treatment, incineration of high calorific fraction sorted in MBP and landfilling of combustion residues.

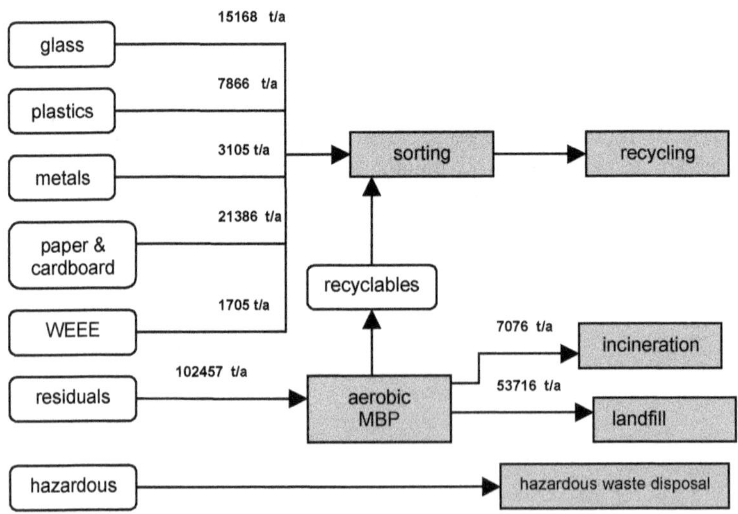

Figure 39. Scenario 4-Kaunas

Outputs from the usage of tools for scenarios-results of modeling

Environmental Assessment

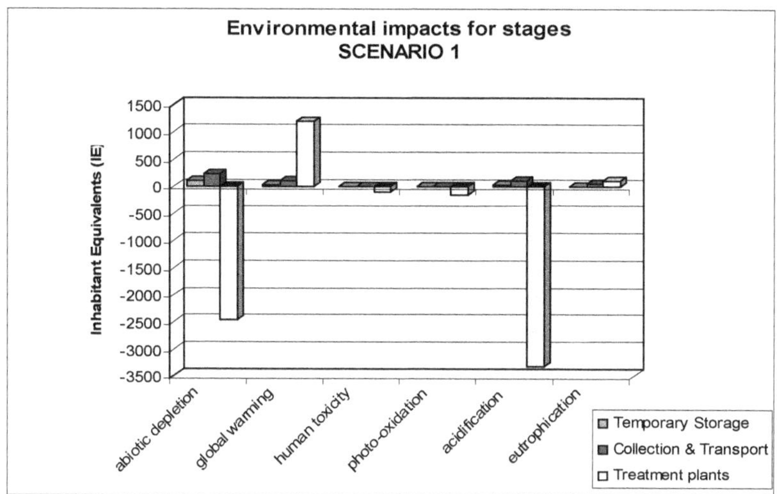

Figure 40. Environmental impacts for subsystems - Kaunas Scenario 1

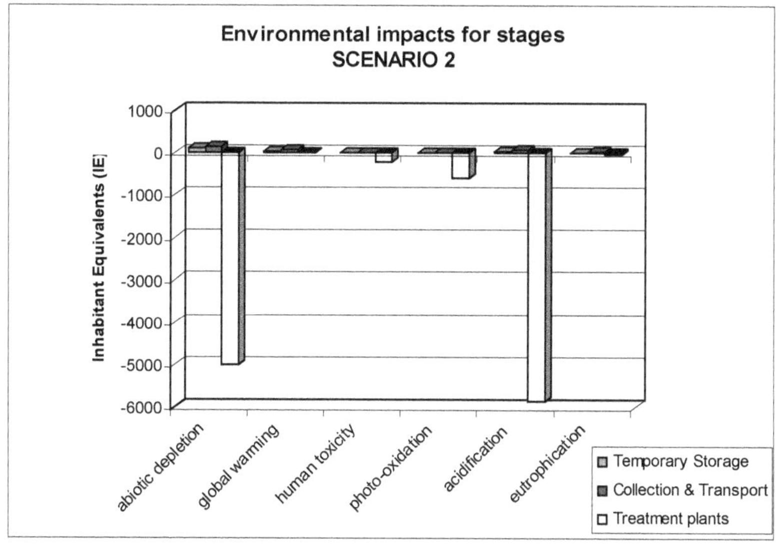

Figure 41. Environmental impacts for subsystems - Kaunas Scenario 2

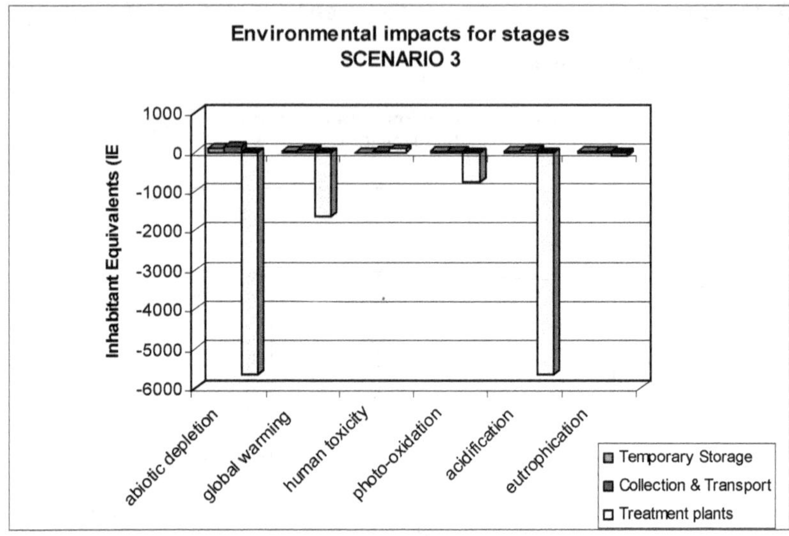

Figure 42. Environmental impacts for subsystems - Kaunas Scenario 3

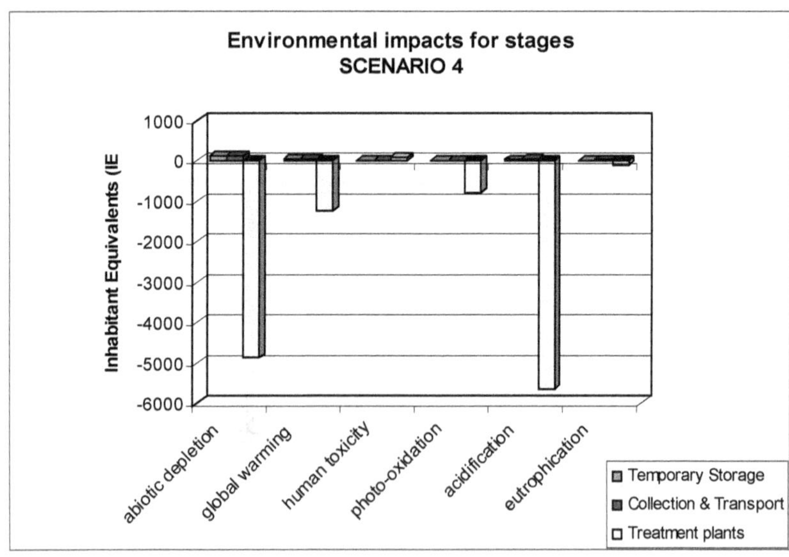

Figure 43. Environmental impacts for subsystems - Kaunas Scenario 4

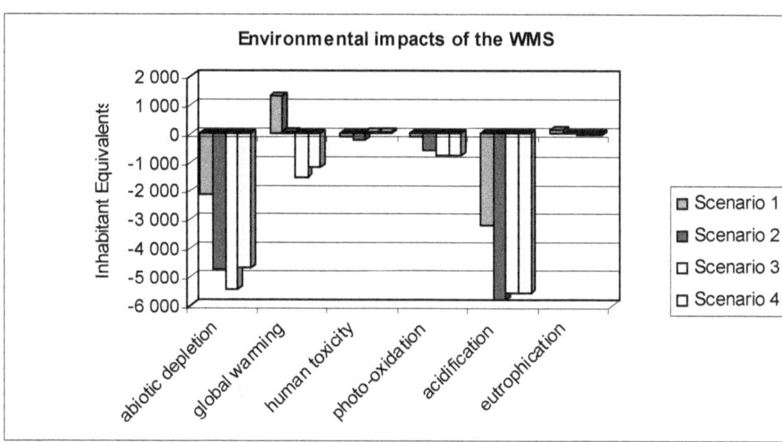

Figure 44. Environmental impacts of MSWMS for Kaunas

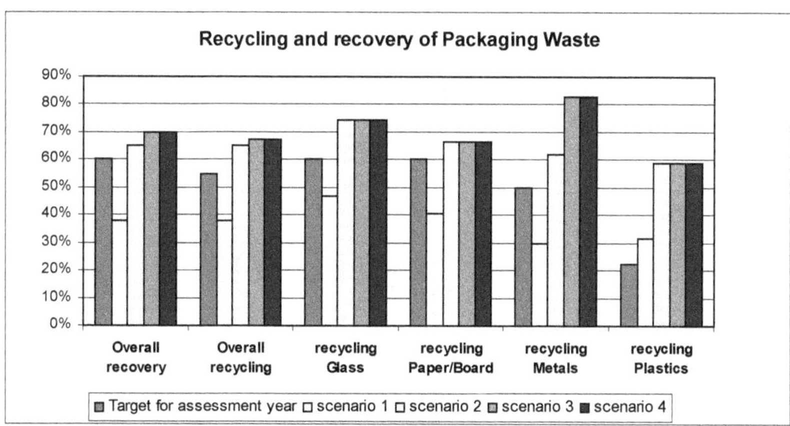

Figure 45. Recycling and recovery of Packaging Waste for Kaunas

Economic Assessment

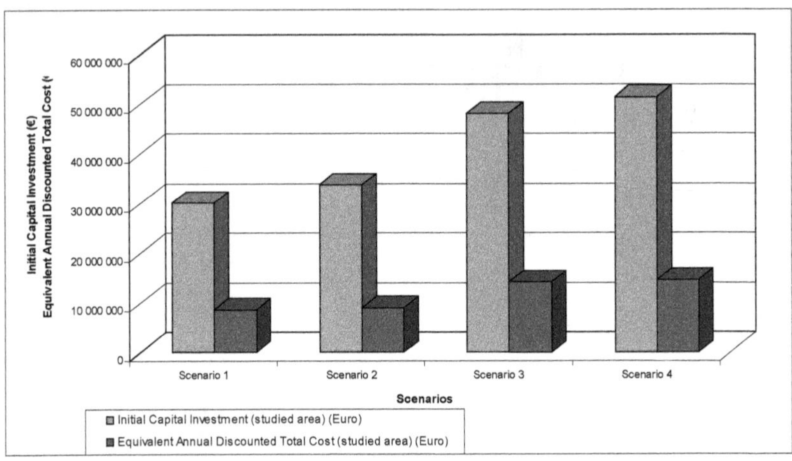

Figure 46. Economic impacts of MSWMS for Kaunas

Table 33. Economic impacts of MSWMS for Kaunas

	Scenario 1	Scenario 2	Scenario 3	Scenario 4
	Initial Capital Investment (€)			
Temp.Storage	11.232.376	15.492.771	15.492.771	15.492.771
Coll. & Transport	8.835.589	6.737.946	6.737.946	6.485.976
Treatment	9.959.449	18.311.674	24.785.725	28.261.869
	Equivalent Annual Capital Cost (€)			
Temp.Storage	1.552.897	2.088.498	2.088.498	2.088.498
Coll. & Transport	851.241	649.149	649.149	624.874
Treatment	6.183.212	6.262.448	11.169.987	11.579.876

Table 34. Economic Efficiency at the municipal level for Kaunas

	Scenario 1	Scenario 2	Scenario 3	Scenario 4
Cost per ton (€/ton)	56,16	58,86	91,08	93,61
Cost per household (€/hh)	60,24	63,13	97,69	100,39
Cost per person (€/person)	22,98	24,08	37,27	38,30
Revenue from recovered material and energy (€)	1.676.680	3.056.273	3.234.748	3.396.760
Total cost as % of GNP of the city (%)	0,51	0,53	0,83	0,85
Diversion between revenue and expenditures for MSWM (%)	37,86	36,12	23,34	22,72

Table 35. Economic Efficiency at the subsystem level (Kaunas)

Subsystem	Cost per ton of waste (€/ton)			
	Scenario 1	Scenario 2	Scenario 3	Scenario 4
Temporary Storage	10	14	14	14
Coll. & Transport	28	23	23	22
Landfill	10	11	11	11
MBP aerobic	0	0	57	57
Incineration	0	0	0	69
Paper sorting	50	49	49	49
Glass sorting	51	51	51	51
Metals sorting	63	62	62	62
Plastics & comp.sort.	27	23	23	23
WEEE sorting	66	61	61	61

Table 36. Equity and Dependence on Subsidies in Kaunas

	Scenario 1	Scenario 2	Scenario 3	Scenario 4
Cost per person as % of minimum wage	370,63	388,45	612,91	629,55
Cost per person / income per person %	0,54	0,57	0,89	0,92
Subsides or grants per person (€/person)	0	0	0	0

Social Assessment

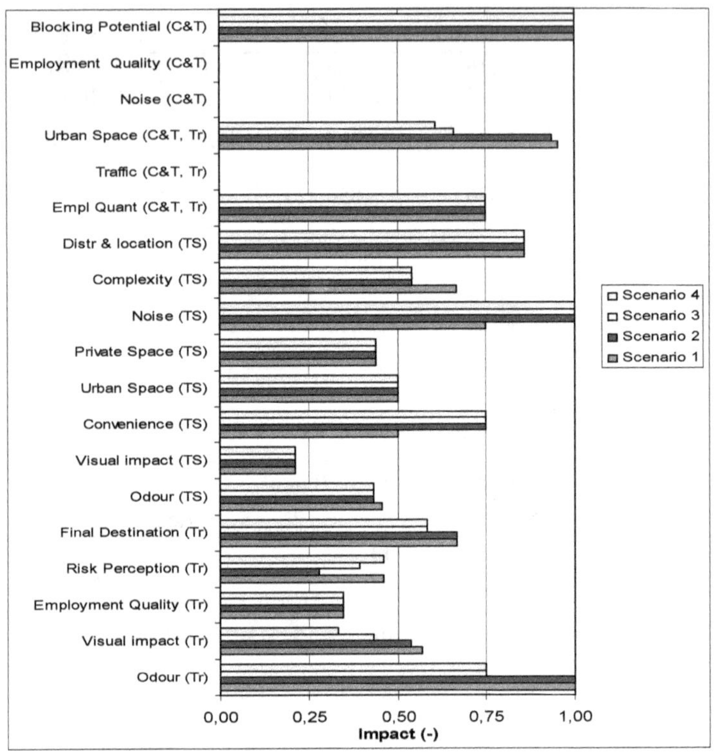

Figure 47. Social impacts of the MSWMS for Kaunas

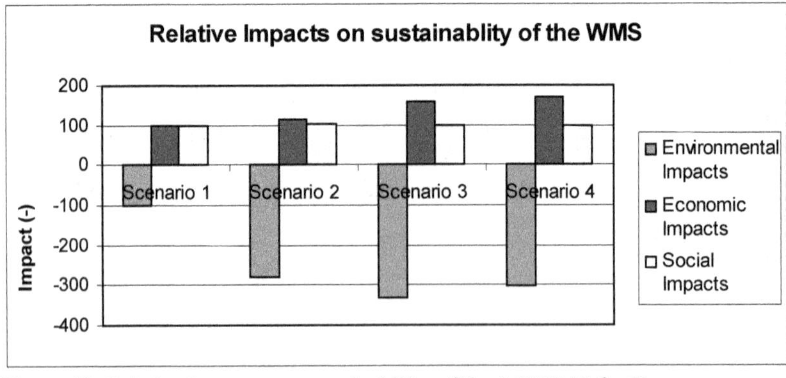

Figure 48. Relative Impacts on sustainability of the MSWMS for Kaunas

Conclusions and recommendations

Since May 2004 Lithuania is a member of the EU. Accordingly the best scenario had to be chosen in accordance with EU legislation. The European Commission has determined a number of specific recovery and recycling targets to be achieved by various target years. The current situation (Scenario 1) does not comply with these targets at all (see Figure 45). This means that the current situation has to be improved. Scenarios 3 and 4 are in compliance with targets for recovery and recycling of packaging waste as well as with targets for diversion of biodegradable waste from landfilling.

In the case of Kaunas city, the most important environmental indicators are global warming, acidification and human toxicity. Regarding these three indicators, the worst scenario is Scenario 1. The Scenarios 3 and 4 show good results in reducing "global warming" and "acidification" (environmental relief), but they also show relatively small environmental burden in impact category "human toxicity".

There is a big difference between investment costs of the scenarios. The investments are €16mio higher in Scenario 3 (for MBP facility) and €20mio higher (for MBP and Incineration facilities) in Scenario 4 comparing with Scenario 1.

The lowest and highest equivalent annual discounted total costs show Scenario 1 (8.587.350 Euro) and Scenario 4 (14.312.451 Euro), respectively. The annual total cost of SWMS vary from 56,16 Euro/t (60,24 Euro/hh, 22,98 €/person) in Scenario 1 to 93,61 Euro/t (100,39 €/hh, 38,30 €/person) in Scenario 4.

The revenue from recovered material and energy vary from €1.676.680 (Scenario 1) to €3.396.760 (Scenario 4).

In Figure 48 relative global impacts of scenarios are compared. They are related to baseline Scenario 1 and regarding social criteria show much lower impacts of Scenarios 3 and 4 on social sustainability of SWMS.

Comparing the above mentioned impacts on environmental, economic and social sustainability of SWMS, Scenario 3 is recommended as the best. Scenario 4 seems to be not realistic due to the very small capacity of incineration plant treating only combustible residues of MBP. There is a need for analysing additional waste treatment options for Kaunas using the developed tool, like incineration of whole waste stream as only treatment option or using biomass as energy source for heating purposes. The composting will be hardly implementable because of the citizen's aversion to separate kitchen waste. According to results of performed studies people are ready to collect plastic and hazardous waste separately.

10.2.3 Wroclaw-Poland

Short characteristics of the city

Wrocław, is located in south-western Poland, on the Odra River. Major east-west and north-south routes intersect in Wrocław: the A-4 motorway (running from Dresden, through Wrocław to Katowice, and in the future to the eastern border of Poland), national ways no 5 and 8 from southern border of Poland leading up-country. Wrocław is also connected to the European system of railways and waterways.

Area: 292,84 km^2. Population: 637,5 thousands inhabitants.

From the early 1990's numerous foreign investors have decided to team up with Polish businesses or purchase existing industrial facilities. Plans for attracting new foreign investment to the city are geared towards the fulfilment of goals defined in the *Wroclaw 2000 Plus* city development strategy

Characteristics of municipal solid waste quantities and composition

Actual municipal solid waste amounts and forecast for Wrocław are presented in the Figure 49. Waste composition in 2004 is shown in Table 37.

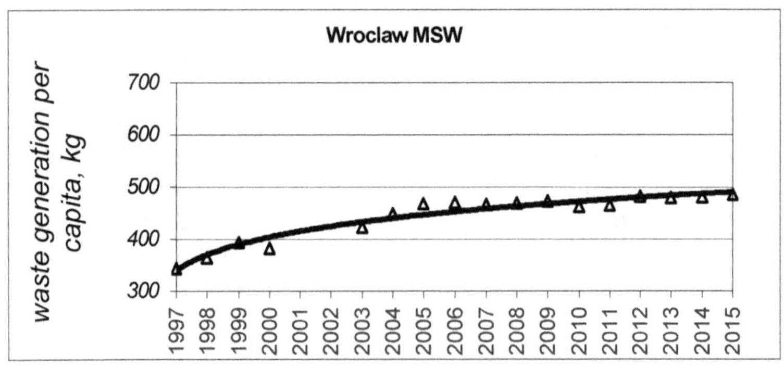

Figure 49. Waste amounts in Wrocław - measured and projected

Table 37 Waste composition in Wroclaw

Year	Organic	Paper & cardboard	Glass	Metals	Plastics	Hazardous	WEEE	Other	Bulky waste
					%				
2004	37,5	9,0	11,0	2,1	11,0	0,2	0,1	29,1	-
2015	33,6	11,4	11,2	1,9	12,0	0,3	0,1	29,5	-

Existing municipal waste management situation

At present, the municipal waste management in Wrocław is limited to only some activities containing:

- temporary storage and collection of mixed waste,
- temporary storage and separate collection of recyclables - mainly glass and plastics,
- collection and transport of mixed waste to transfer stations situated in the city or directly to landfill sites located outside Wrocław in various distances from the city center - from ca 36km to ca 90km and
- collection and transport of separately collected materials to sorting plant, sorting, baling and preparing for transport to recycling plants.

Wrocław as a municipality does not have any waste treatment plant and any landfill. There is no municipal company operating the waste management system of the city. Municipality has given the permits (licences) for waste collection and transport to more than 50 companies, among which only a few are really providing services for citizens and business.

Transfer station serves only a part of the waste stream generated in Wrocław - it was ca 160.000 tons in 2002, about 70 % of the waste stream. Other firms operate simplified unloading of waste from collection vehicles to big-size containers without transfer station.

Figure 50 presents a flow chart for existing waste management situation in Wroclaw as baseline scenario 1.

Scenario 1: Separate collection and sorting of glass, plastics & metals and paper & cardboard for recycling, transport of residuals *via* transfer station to landfill.

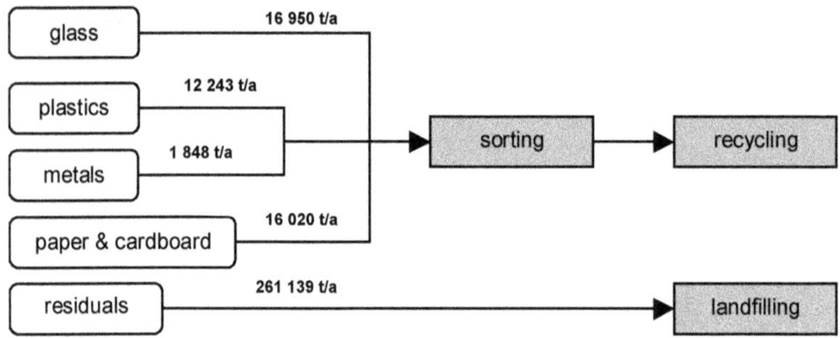

Figure 50. Current waste management in Wroclaw (Scenario 1)

Description of chosen waste management scenarios for Wroclaw.

Scenario 2: Separate collection and sorting of glass, MDR (plastics & metals, paper & cardboard), WEEE and bulky waste for recycling, hazardous waste for treatment, green waste (from public green area) for composting in windrows, aerobic mechanical-biological pre-treatment of residual waste before landfilling.

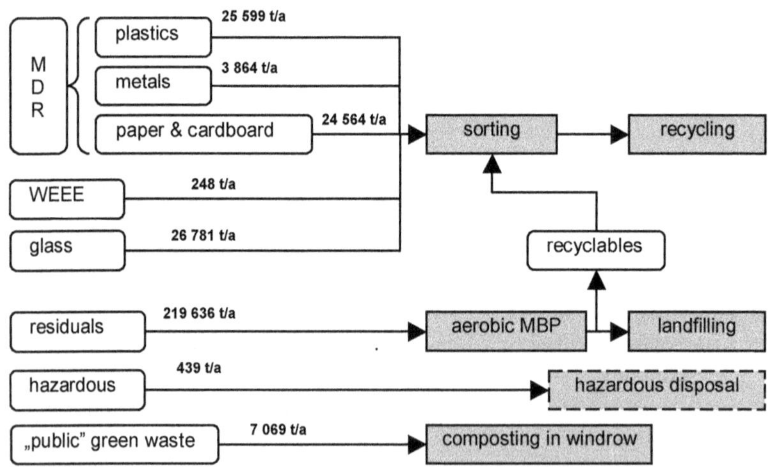

Figure 51. Wroclaw – Scenario 2

Scenario 3: Separate collection and sorting of glass, plastics & metals, paper & cardboard, bulky waste and WEEE for recycling, hazardous waste for treatment, bio-waste (kitchen and garden) and 'public' green waste for composting in facility, aerobic mechanical-biological pre-treatment of residual waste before landfilling with separation of High Caloric Fraction and usage as RDF in cement industry.

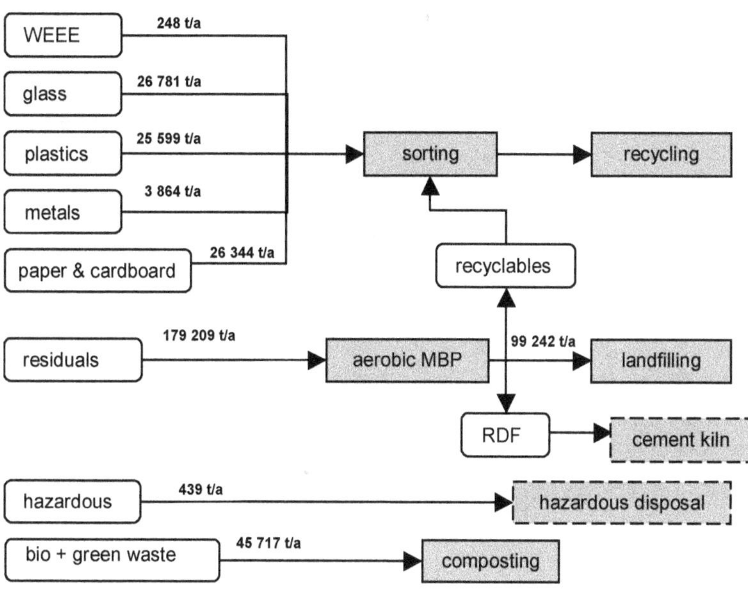

Figure 52. Wroclaw – Scenario 3

Scenario 4: Separate collection and sorting of glass, MDR (plastics & metals, paper & cardboard), WEEE and bulky waste for recycling, hazardous waste for treatment, bio-waste (kitchen and garden) and 'public' green waste for composting in facility, transfer station for residual waste, incineration of residual waste, landfilling of incineration residues.

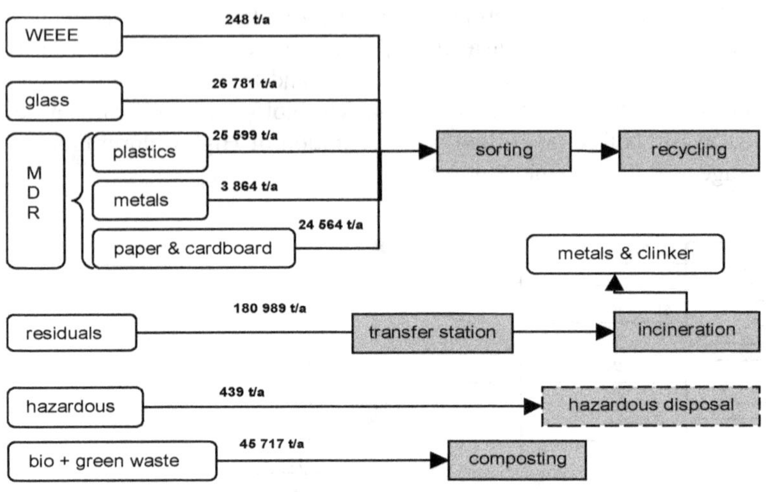

Figure 53. Wroclaw - Scenario 4

Outputs from the usage of tools for scenarios- results of modeling

Environmental Assessment

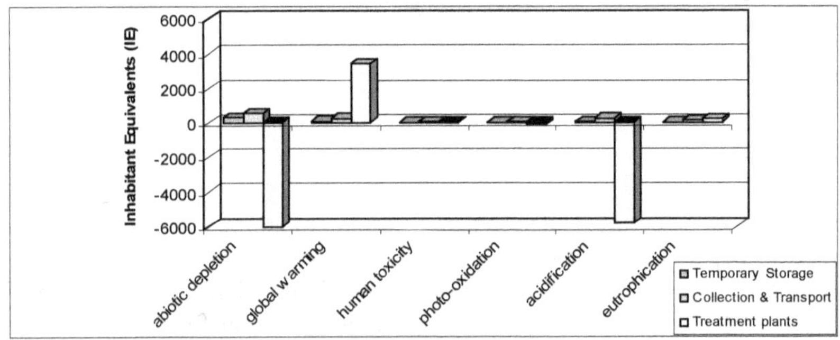

Figure 54. Environmental impacts of subsystems - Wroclaw Scenario 1

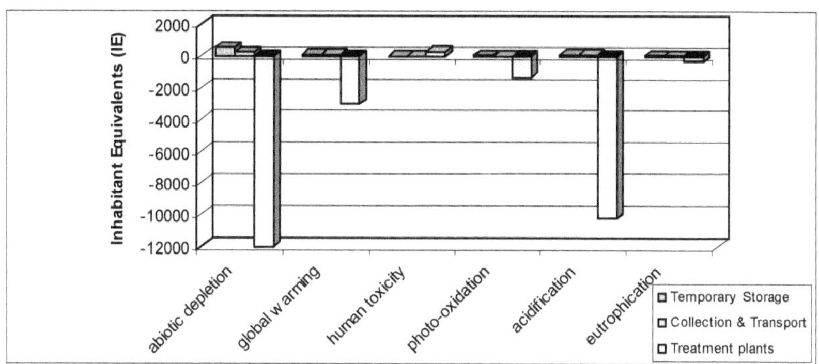

Figure 55. Environmental impacts of subsystems - Wroclaw Scenario 2

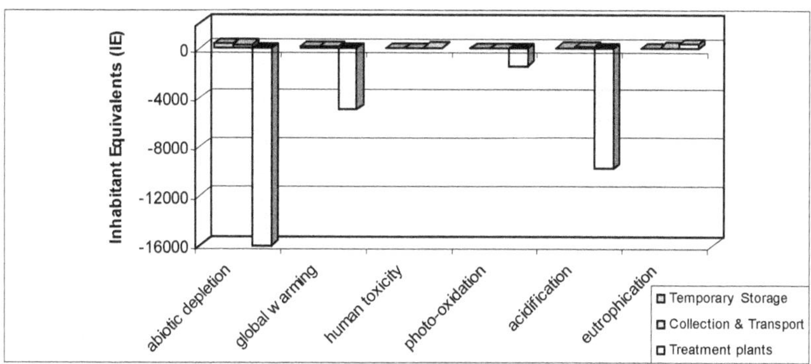

Figure 56. Environmental impacts of subsystems - Wroclaw Scenario 3

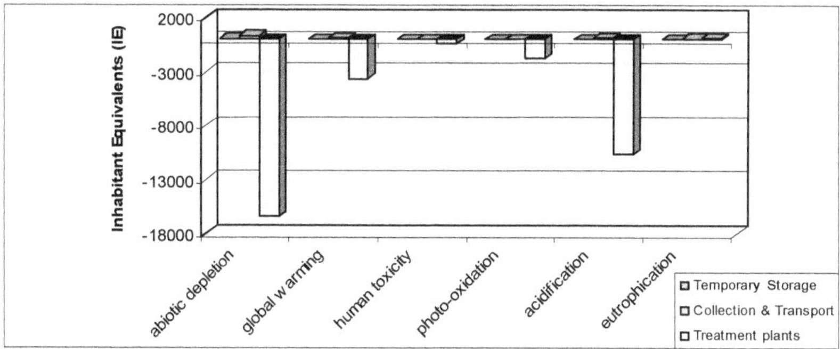

Figure 57. Environmental impacts of subsystems - Wroclaw Scenario 4

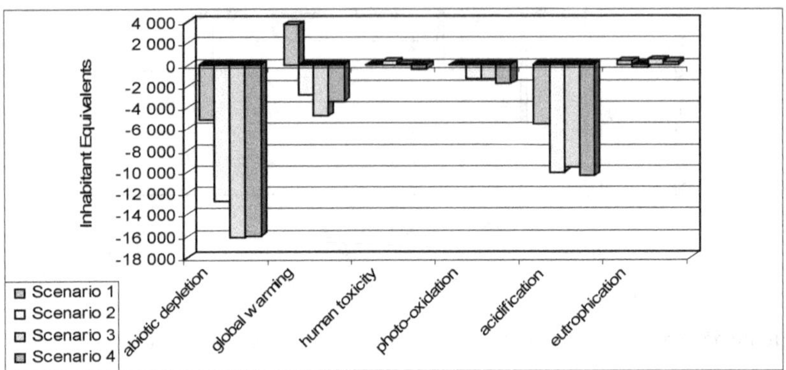

Figure 58. Environmental impacts of MSWMS for Wroclaw

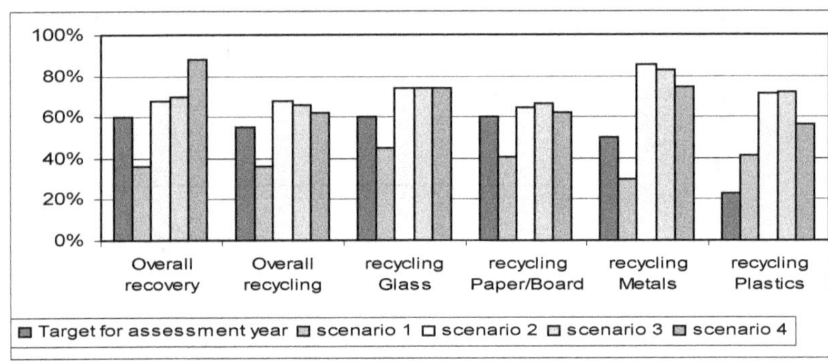

Figure 59. Recycling and recovery of Packaging Waste for Wroclaw

Economic Assessment

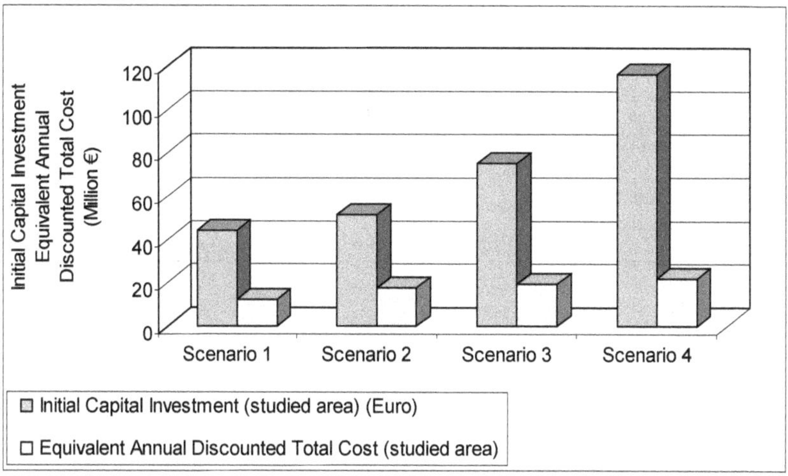

Figure 60. Economic impacts of MSWMS in Wroclaw

Table 38. Economic impacts of MSWMS in Wroclaw

	Scenario 1	Scenario 2	Scenario 3	Scenario 4
	Initial Capital Investment (10^6 €)			
Temp.Storage	7,9	5,3	9,0	5,5
Coll. & Transport	14,2	7,9	12,9	10,3
Treatment	22,1	37,9	53,3	100,7
	Equivalent Annual Discounted Total Cost (10^6 €)			
Temp.Storage	1,8	2,0	2,1	1,2
Coll. & Transport	6,4	3,5	5,5	4,6
Treatment	3,7	12,0	12,2	16,1

Table 39. Economic Efficiency at the municipal level in Wroclaw

	Scenario 1	Scenario 2	Scenario 3	Scenario 4
Cost per ton (€/ton)	38,55	57,88	64,09	71,11
Cost per household (€/hh)	52,46	78,76	87,21	96,77
Cost per person (€/person)	18,64	27,98	30,99	34,38
Revenue from recovered material and energy (€)	2 917 018	4 626 673	6 523 312	8 117 456
Total cost as % of GNP of the city (%)	0,34%	0,50%	0,56%	0,62%
Diversion between revenue and expenditures for MSWMS (%)	134,14%	89,35%	80,68%	72,72%

Table 40. Economic Efficiency at the subsystem level in Wroclaw

Subsystem	Cost per ton of waste (€/ton)			
	Scenario 1	Scenario 2	Scenario 3	Scenario 4
Temporary Storage	5,89	6,42	6,68	4,03
Coll. & Transport	20,78	11,47	17,73	14,93
Landfill	10,73	8,37	17,13	26,60
Composting	0,00	50,74	30,66	29,84
Digestion	0,00	0,00	0,00	0,00
MBP aerobic	0,00	40,56	43,37	0,00
MBP anaerobic	0,00	0,00	0,00	0,00
Incineration	0,00	0,00	0,00	67,62
Paper sorting	25,67	0,00	28,25	0,00
Glass sorting	14,61	14,03	16,02	15,72
Metals sorting	29,82	0,00	19,30	0,00
Plastics & comp.sort.	7,41	0,00	3,55	0,00
MDR sorting	0,00	33,04	0,00	33,04
WEEE sorting	0,00	65,78	70,00	69,80

Table 41. Equity and Dependence on Subsidies for Wroclaw

	Scenario 1	Scenario 2	Scenario 3	Scenario 4
Cost per person as % of minimum wage	196,18%	294,53%	326,16%	361,90%
Cost per person / income per person	0,27%	0,41%	0,45%	0,50%
Subsides or grants per person (€/person)	5,88	5,67	8,60	11,42

Social Assessment

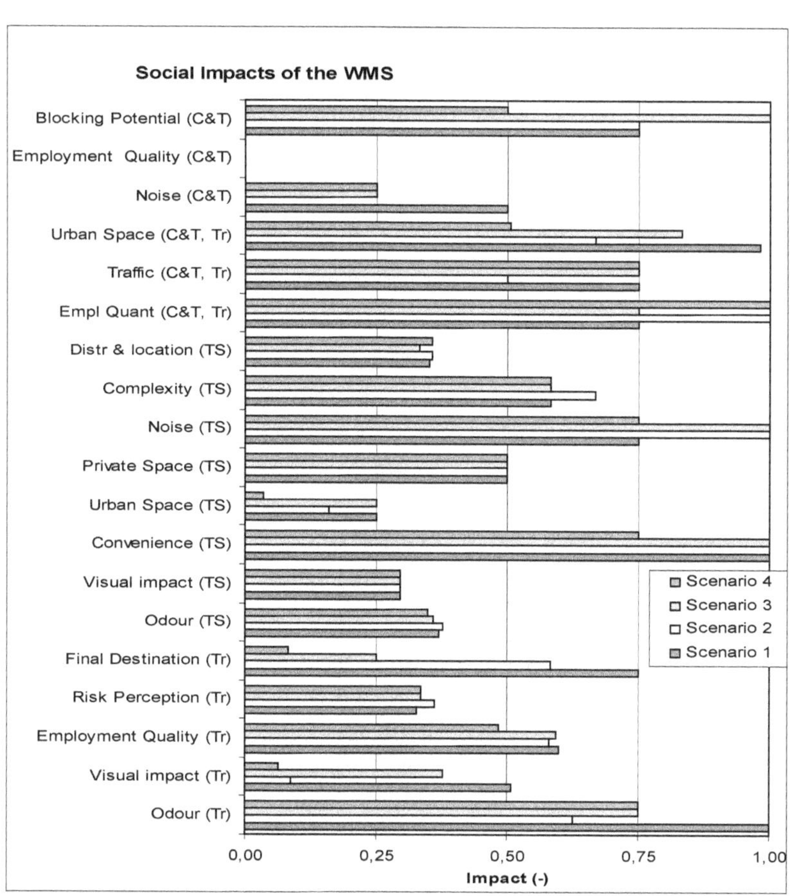

Figure 61 Social impacts of the MSWMS in Wroclaw

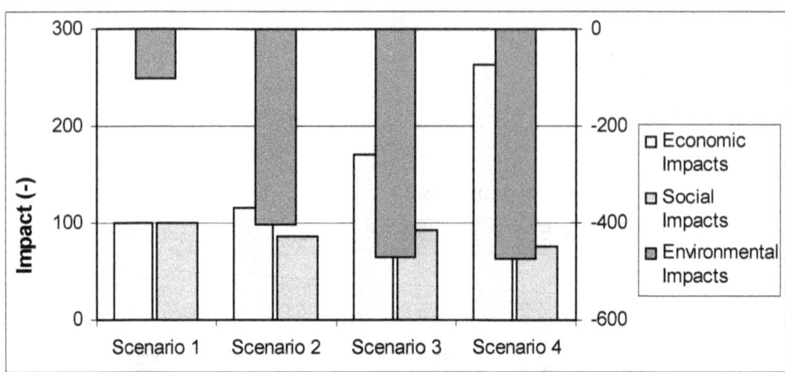

Figure 62 Relative Impacts on sustainability of the MSWMS in Wroclaw

Conclusions and recommendations

Base scenario 1, containing the present situation of waste management in Wroclaw, was compared with three possible scenarios of future Wroclaw's MSWMS. Different ways of treatment of main residual waste streams (aerobic MBP or incineration) and different ways of separate collection were proposed. The comparison of environmental impacts of subsystems, in particular scenarios, shows that waste treatment in scenario 1 (present situation) causes the absolutely highest environmental burden (in the impact category global warming) among all impact categories in all scenarios. The environmental burdens of other MSWMS stages in scenarios 2-4 are rather low in comparison to that one above mentioned.

Condensed comparison of impacts of scenarios on sustainability of MSWMS is shown in Figure 62. Relative impacts of scenario 1 are given a value of 100%, to which the other scenarios are compared.

Intensified extent of waste treatment (and minimizing waste landfilling) leads to overall growing environmental relief in scenarios 2 to 4 (negative impacts). It means that credits from recovery of materials and energy, even in present situation of MSWMS, surpass environmental burdens of some subsystems. Only impact categories eutrophication and human toxicity show relatively small environmental burden in some scenarios. Scenario 4 entails the highest environmetal relief among all scenarios assessed.

Economic and social impacts take positive values. Investment costs of MSWMS proposed in scenario 2 are ca. 15% bigger than the costs of the present system (scenario 1), while investments proposed in scenario 3 and 4 are ca. 70% and almost three time more expensive in comparison to scenario 1, respectively.

Annual cost of MSWMS in scenario 2 is ca. 50%, in scenario 2 – ca. 66% and in scenario 4 – ca. 84% higher than at present (scenario 1).

All proposed scenarios should be socially acceptable. Scenarios 2-4 are characterised by improved social perception of MSWMS.

All scenarios of future MSWMS achieve targets of Landfill and Packaging Directives.

Comparing relative impacts of all proposed scenarios, scenario 3 seems to combine relatively high environmental profits with moderate costs and high social acceptability. Scenario 4, containing incineration, is even little more socially accepted, but provides better environmental profits, but require double the investment.

10.2.4 Nitra-Slovakia

Short characteristics of the city

Nitra is the oldest Slovakian city situated in western part of Slovakia with a beautiful panorama of Zobor hill. Zobor belongs to Tribec mountains and all to Carpathian mountains. The splendid scenery is enhanced by a meandering river of the same name and several little hills. Geographical coordinates are as follows: latitude 48°15′ and longitude 18°10′. The average yearly temperature is nearly 10° C with July being the warmest with average temperatures about 20° C and January being coldest with the average temperature slightly below zero.

Nitra is the is administrative centre of the Nitra region, which compose districts of Nitra, Komárno, Levice, Nové Zámky, Šaľa, Topoľčany a Zlaté Moravce.

Today Nitra is the seat of science, education, economy, culture, clergy and sport. Nitra region with exhibition ground of the Agrokomplex is interna-tional centre of exhibitions. Nitra is a modern city of young people; there are two Universities: the Slovak Agriculture University and the University of Constantine, the Philosopher.

Total area 108 km2 - Population 87,285 inh /26.5.2001/.

Characteristics of municipal solid waste quantities and composition

Actual municipal solid waste amounts and forecast for Nitra are shown in Table 42. Waste composition in 2004 and 2015 is presented in Table 43.

Table 42. Municipal Solid Waste actual and projected amounts - Nitra

Year	2001	2002	2003	2004	2005	2006	2007	2008	2009	2010	2011	2012	2013	2014	2015
t/a	33 623	32 080	34 056	33 909	34 535	35 105	35 686	36 288	36 908	37 542	38 192	38 859	39 542	41 251	43 143

Table 43. Waste composition in Nitra - actual data and forecast, mass %

Year	Organic	Paper & cardboard	Plastics , composites	Glass	WEEE	Metals	Hazardous	Other	Bulky waste
					%				
2004	31.5	18.6	8	8.1	1.7	3.5	0.9	12.4	13.3
2015	27.6	19.3	8	8.5	1.8	3.5	0.9	13	14.1

Existing municipal waste management situation

Collection and transport of solid waste in residential areas in Nitra is covered by three systems:
A - waste collection using conventional vehicles: KUKA , BOBR, Linear (internal compactor), types of containers - 110, 120, 240, 1100 l,
B - waste collection with system L.S.L. PACKER (self-loading with internal compactor, types of containers - 120, 240, 1100 l,
C - waste collection with 'Boom – arm – loader – dumptruck'.

Transfer station

Transfer station is a simple construction - reinforced concrete unloading platform with 2 systems: 1- waste unloading from collection vehicles

KUKA, BOBR, Linear, and 2 – unloading from L.S.L PACKER. Containers are unloaded from the collection vehicles to the truck and pull trailer.

Landfill

Landfill is situated in an area in the village of Nový Tekov - 42 km from Nitra.It was designed in 1994 with the start of the landfill operation in 1996 and an operational time to 2044.

Leachate management - leachate collection system (perforated collection pipe, washed gravel) - storm retention basin - leachate recycling - a part treatment in waste water treatment plan.

Landfill gas management - passive control of landfill gas with gas vent well.

Separate collection

Separate collection of paper, glass, metal, accumulators, plastic, fluorescent tubes, waste oil, garden waste is operated in the city.

Composting

Composting yard waste – windrow system.

Pretented is concept of garden waste pretreatment by shredding with subsequent composting in open static piles.

Flow chart of existing waste management situation in Nitra is shown in Figure 63.

Description of chosen waste management scenarios for Nitra.

Scenario 1: Separate collection of recyclables (paper and cardboard, glass, plastics), composting of garden waste, landfilling of residual waste.

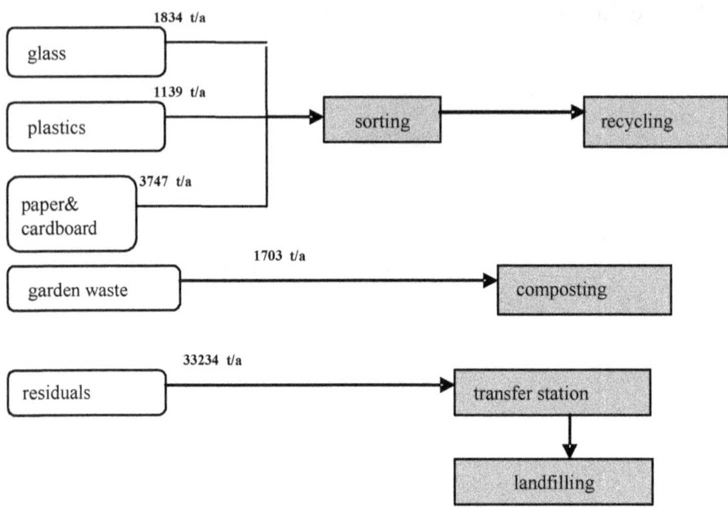

Figure 63. Current waste management in Nitra (Scenario 1)

Scenario 2: Separate collection of recyclables (glass, plastics, paper and cardboard, metals), aerobic mechanical biological pre-treatment, landfilling of the low calorific fraction, incineration of the high calorific fraction.

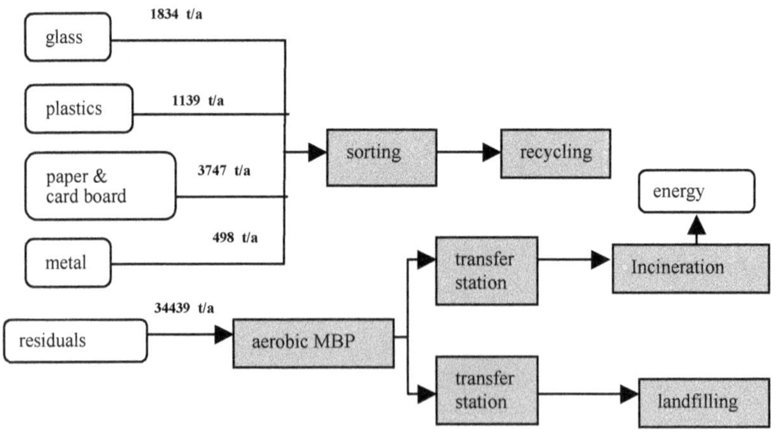

Figure 64. Nitra – Scenario 2

Scenario 3 Separate collection of recyclables (glass, plastics, paper and cardboard, metals), composting of garden waste, aerobic mechanical- biological pretreatment before landfilling

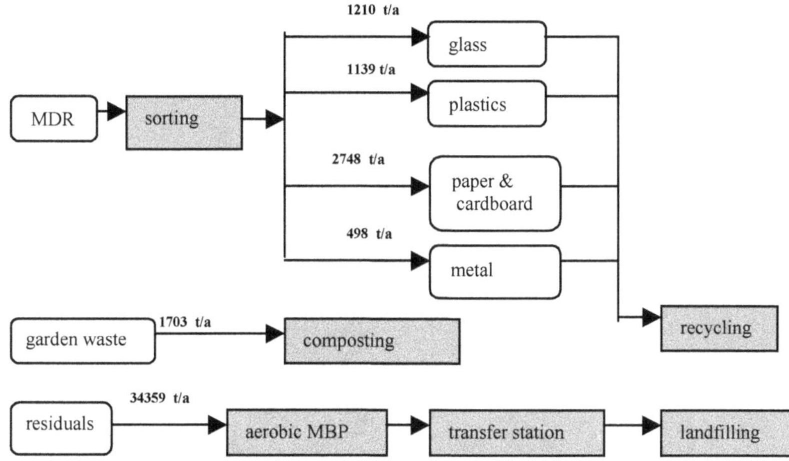

Figure 65. Nitra - Scenario 3

Scenario 4 - Separate collection of recyclables (plastic, paper and cardboard, metals, glass, and WEEE), aerobic mechanical-biological pretreatment, landfilling.

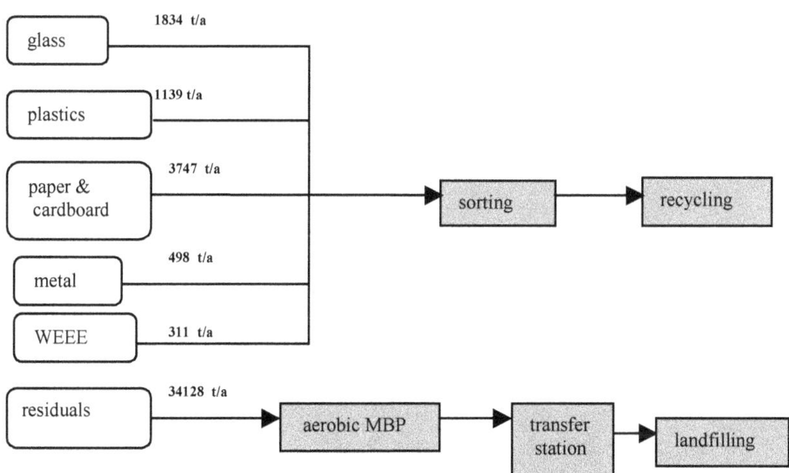

Figure 66. Nitra - Scenario 4

Outputs from the usage of tools for scenarios- results of modeling

Environmental Assessment

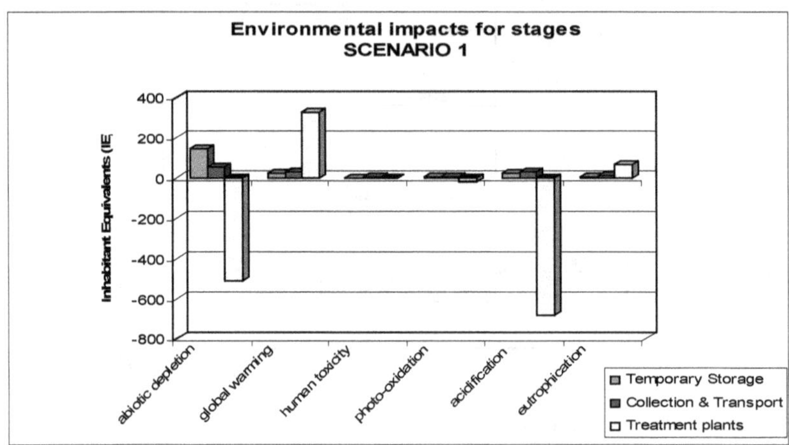

Figure 67. Environmental impacts for subsystems – Nitra Scenario 1

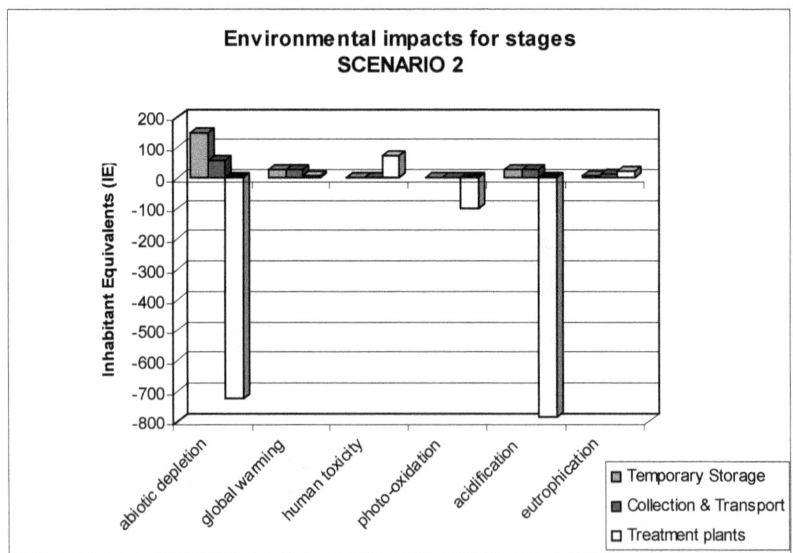

Figure 68. Environmental impacts for subsystems - Nitra Scenario 2

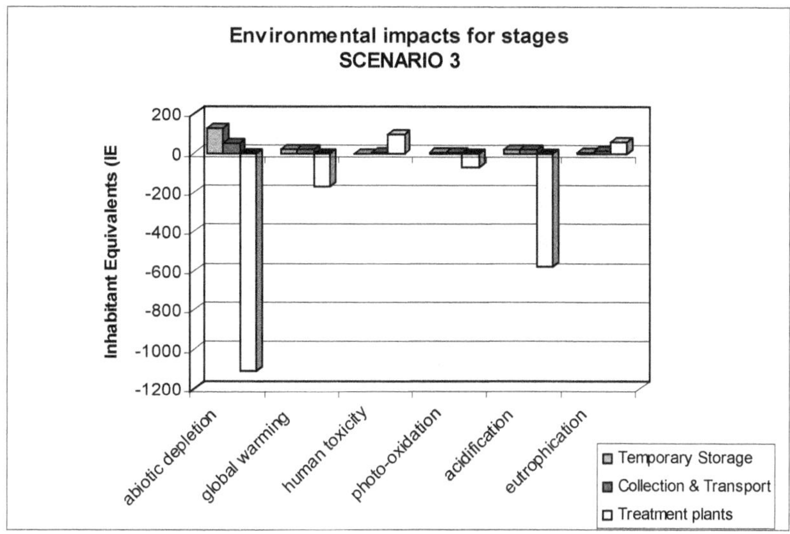

Figure 69. Environmental impacts for subsystems - Nitra Scenario 3

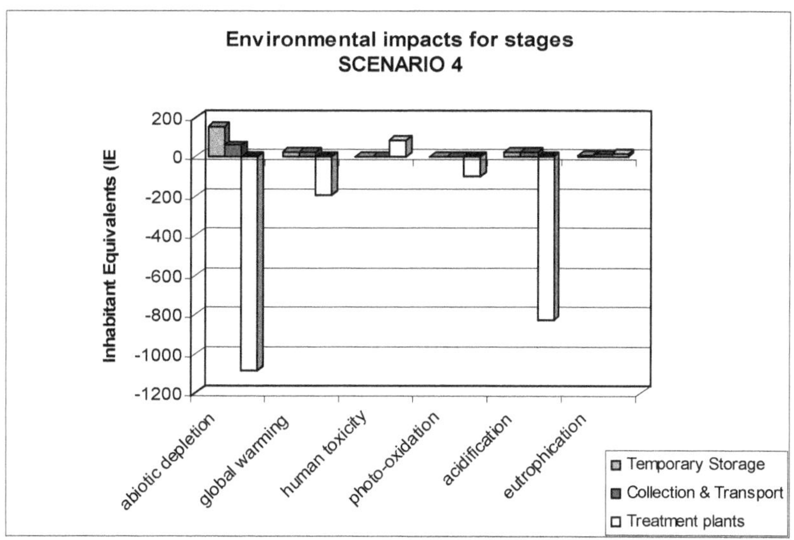

Figure 70. Environmental impacts for subsystems - Nitra Scenario 4

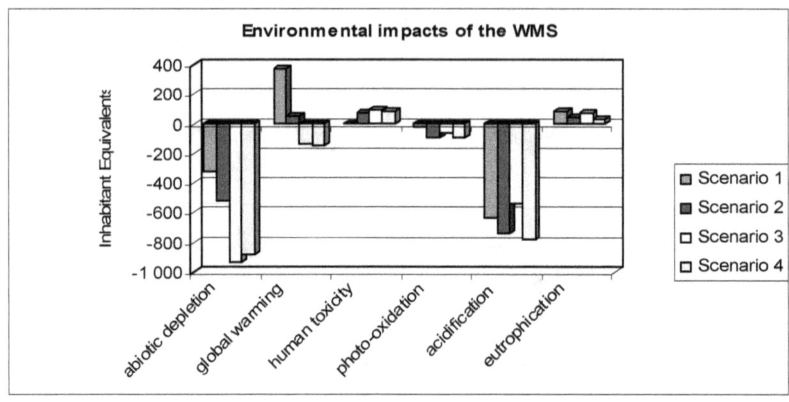

Figure 71. Environmental impacts of MSWMS in Nitra

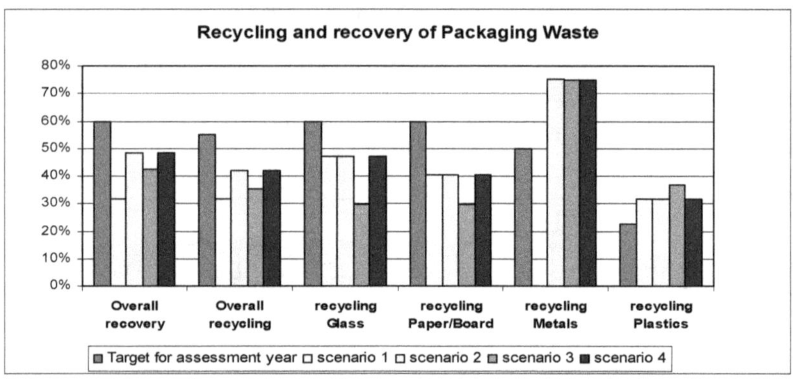

Figure 72. Recycling and Recovery of Packaging Waste for Nitra

Economic Assessment

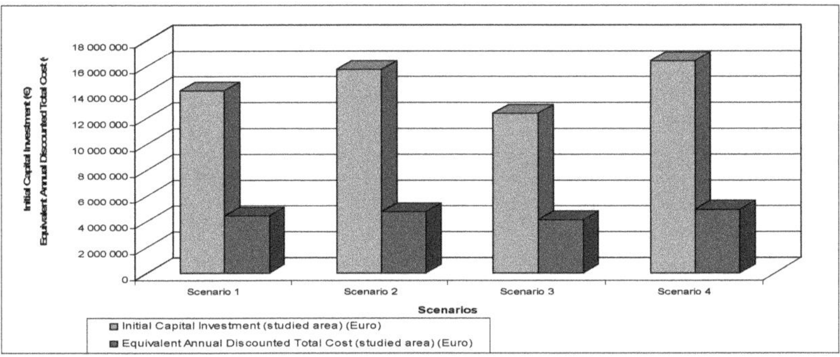

Figure 73. Economic impacts of MSWMS for Nitra

Table 44. Economic impacts of MSWMS in Nitra

	Scenario 1	Scenario 2	Scenario 3	Scenario 4
	Initial Capital Investment (10^6 €)			
Temp.Storage	6,0	6,3	5,8	6,3
Coll. & Transport	2,3	2,4	2,2	2,4
Treatment	5,8	6,9	4,3	7,7
	Equivalent Annual Capital Cost (10^6 €)			
Temp.Storage	1,4	1,5	1,3	1,5
Coll. & Transport	1,1	1,2	1,1	1,1
Treatment	1,9	2	1,7	2,3

Table 45. Economic Efficiency at the municipal level in Nitra

	Scenario 1	Scenario 2	Scenario 3	Scenario 4
Cost per ton (€/ton)	105,81	114,53	98,55	118,97
Cost per household (€/hh)	119,94	129,08	111,70	134,07
Cost per person (€/person)	50,77	54,63	47,28	56,75
Revenue from recovered material and energy (€)	222.478	496.995	360.713	470.999
Total cost as % of GNP of the city (%)	1,24%	1,34%	1,16%	1,39%
Diversion between revenue and expenditures for MSWMS (%)	49,24%	45,76%	52,88%	44,05%

Table 46. Economic Efficiency at the subsystem level in Nitra

Subsystem	Cost per ton of waste (€/ton)			
	Scenario 1	Scenario 2	Scenario 3	Scenario 4
Temporary Storage	34	36	33	36
Coll. & Transport	27	28	25	28
Composting	30	16	33	34
Landfill	62	0	62	0
Digestion	0	0	0	0
MBP aerobic	0	23	18	23
MBP anaerobic	0	0	0	0
Incineration	0	63	0	0
Paper sorting	93	93	0	93
Glass sorting	98	98	0	98
Metals sorting	0	118	0	118
Plastics & comp.sort.	203	203	0	203
MDR sorting	0	0	57	0
WEEE sorting	0	0	0	122

Table 47. Equity and Dependence on Subsidies for Nitra

	Scenario 1	Scenario 2	Scenario 3	Scenario 4
Cost per person as % of minimum wage	507,67%	546,35%	472,80%	567,50%
Cost per person / income per person	1,49%	1,61%	1,39%	1,67%
Subsides or grants per person (€/person)	17,77	19,28	16,47	19,67

Social Assessment

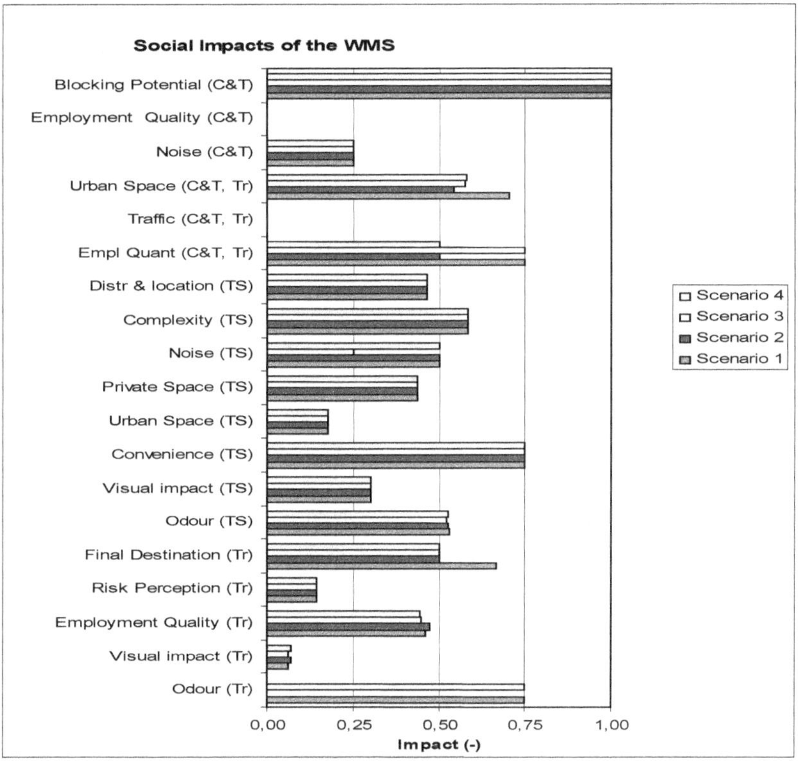

Figure 74. Social impacts of the MSWMS for Nitra

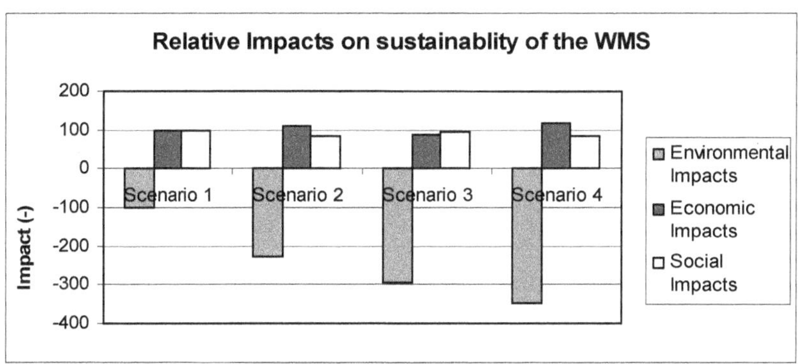

Figure 75.Relative Impacts on sustainability of the MSWMS for Nitra

Conclusions and recommendations

Baseline scenario 1, describing the present situation of waste management in Nitra, was compared with three possible scenarios of future Nitra's MSWMS. Different ways of treatment of main waste stream (aerobic MBP or incineration) and various alternatives of separate collection were proposed.

Comparison of environmental impacts of MSWMS stages (Figure 71) shows that Temporary Storage and Collection & Transport contribute mainly to the impact categories: abiotic depletion, global warming, photo-oxidant formation and acidification while treatment stage influences only human toxity and eutrophication. This observation concerns new scenarios 2-4. Because of credits allocated to the treatment stage of MSWMS (recovery of materials and energy) there is visible environmental relief concerning impact categories abiotic depletion and acidifiction in all scenarios (1-4), global warming in scenarios 3 and 4 as well as photo-oxidant formation in scenarios 2-4. Total relative environmental impacts for each scenario assessed (Figure 71) take negative values what means that positive aspects of material and energy recovery in treatment stage, even in present waste management situation, surpass environmental burdens. In this respect the most advantageous is scenario 4.

Economic and social relative impacts take positive values. In comparison to scenario 1, all proposed new scenarios 2-4 present better social acceptability, especially scenarios 2 and 4. However, development of scenarios 2 and 4 entails higher economic impacts in comparison to scenario 1 while scenario 3 is even more effective economically than the present situation (scenario 1).

Taking into account all relative impacts of scenarios on sustainability of MSWMS it is visible that scenario 3 combines relatively high environmental profits with the lowest costs and high social acceptability. It is recommended for implementation.

10.2.5 Reus-Spain

Short characteristics of the city

The town of Reus rises up in the centre of a coastal plain, in a privileged enclave within southern Catalonia, in Spain. The direct influence of the

Mediterranean and the protection of the nearby mountains make the climate benign all year round, with an average temperature of 17 ° C and not excessively humid. The area of the city is 53 km^2 and the population is approximately 98.000.

Reus is an important center of economy and science. In recent years Reus has made a considerable effort to move with the times and offers its citizens wider and more comprehensive services while maintaining the town's role as a point of reference for other towns and villages.

Characteristics of municipal solid waste quantities and composition

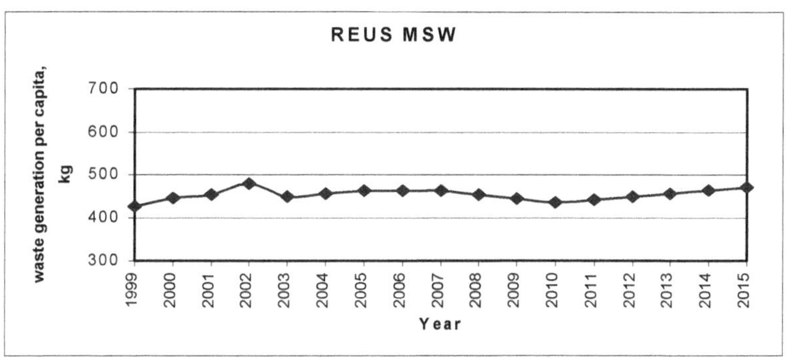

Figure 76. Municipal Solid Waste (MSW) actual and forecasted amounts – Reus

Table 48. Waste composition in Reus - actual data and forecast, mass %

Year	Organic	Paper & cardboard	Glass	Metals	Plastics	Hazardous	WEEE	Other	Bulky waste
					%				
2003	32.5	23.6	9.2	3.6	12.2	1.0	2.0	13.9	2.1
2013	30.2	23.5	9.7	3.7	13.1	1.0	2.1	14.6	2.2

Existing municipal waste management situation

Reus makes separate collection of bulky wastes since 1985, paper/cardboard and glass since 1994, packaging and mixed dry recyclables since 1998, garden and hazardous waste since 1999, bio-waste since 2000 and metals and WEEE since 2003. Residual waste is treated by incineration and the ashes are sent to hazardous waste landfill.

Bio-waste is composted to produce compost. Bio-waste amounts to 23% of the total mass of waste separately collected.

Paper, glass and other recyclables are recycled in specialised facilities to produce secondary materials. Paper is of 32% of total waste of separation collection, glass is 7,5% and packaging is 10%.

Flow chart of present waste management of Reus is shown in Figure 77.

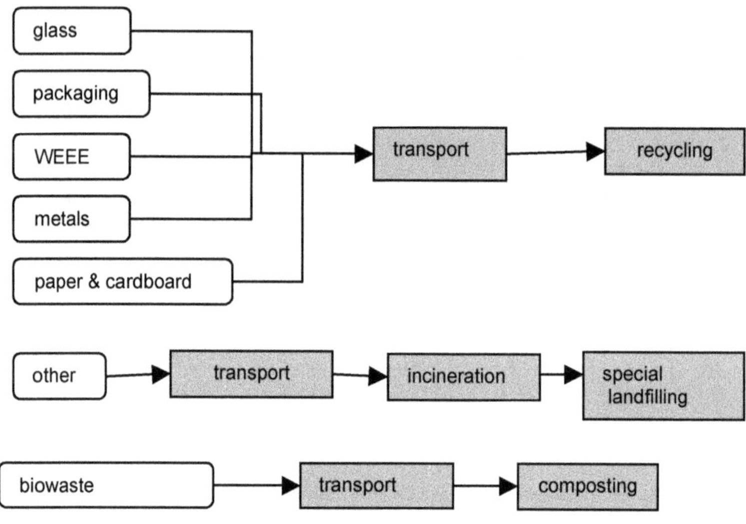

Figure 77. Actual waste management in Reus

Description of chosen waste management scenarios for Reus.

Actual waste management strategy of the municipality of Reus (Scenario 3) is quite advanced and well implemented. So, unlike other cities, two selected scenarios represent past situations (scenario 1 and 2) and another one looks into a hypothetic future (scenario 4). The two scenarios reflecting past situations were very different one to another and the fourth one was similar to the actual one.

Scenario 1: Landfilling of all waste streams.

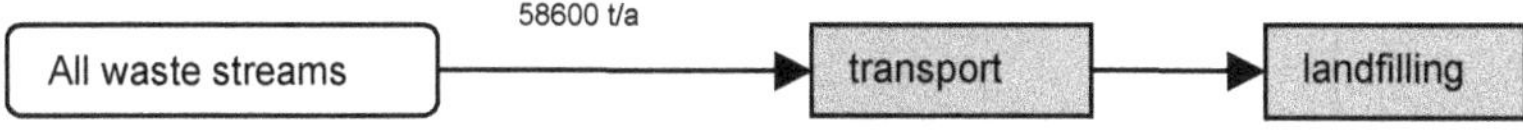

Figure 78. Reus – Scenario 1

Scenario 2: Separate collection of recyclables (glass, plastic, paper & cardboard, metal, WEEE), incineration of residual waste followed by special landfilling.

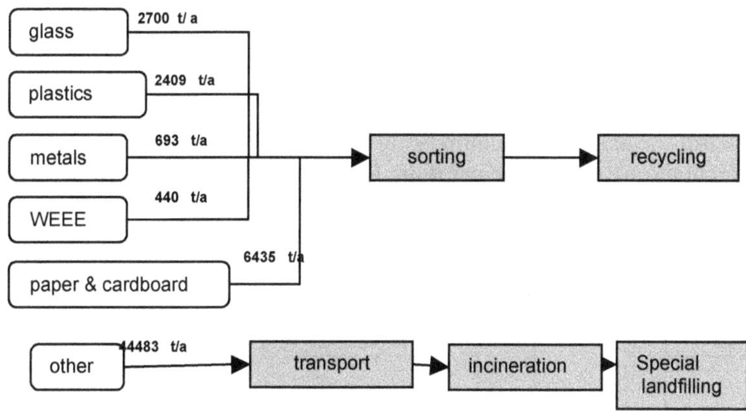

Figure 79. Reus – Scenario 2

Scenario 3: Separate collection of recyclables (glass, plastic, paper & cardboard, metal, WEEE), composting of biowaste, incineration of the residual waste followed by special landfilling.

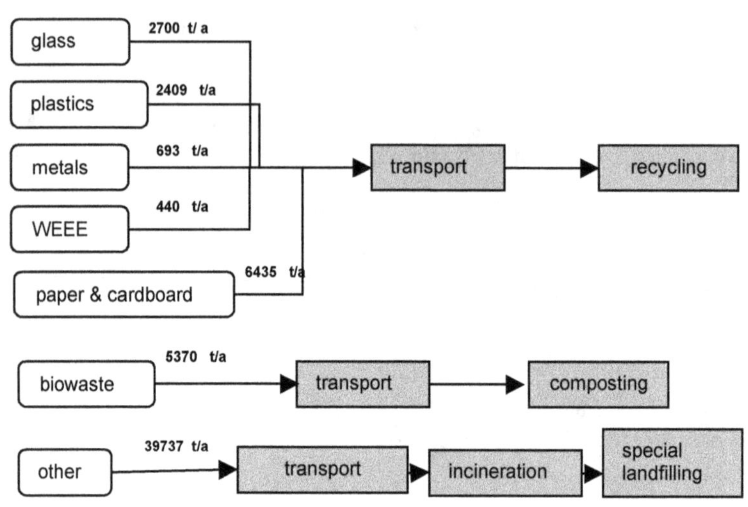

Figure 80. Reus – Scenario 3

Scenario 4: Separate collection of recyclables (glass, plastic, paper & cardboard, metal, WEEE), composting of biowaste, anaerobic mechanical-biological pre-treatment, incineration, followed by special landfilling.

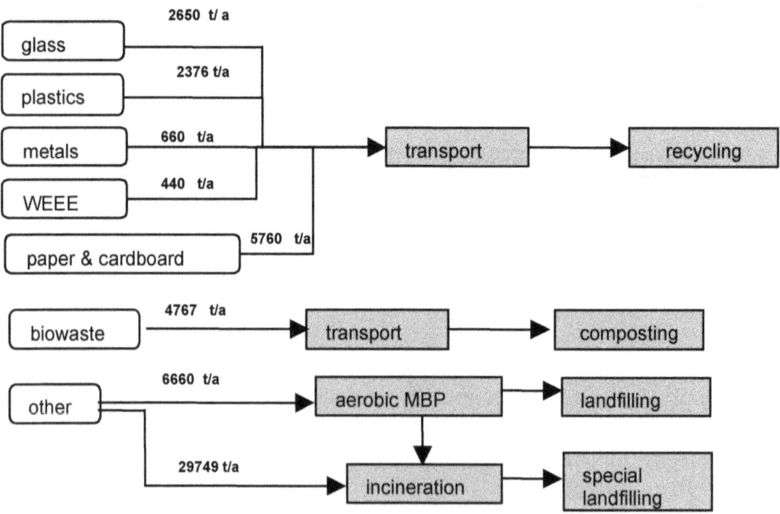

Figure 81.Reus-scenario 4

Outputs from the usage of tools for scenarios- results of modeling

Environmental Assessment

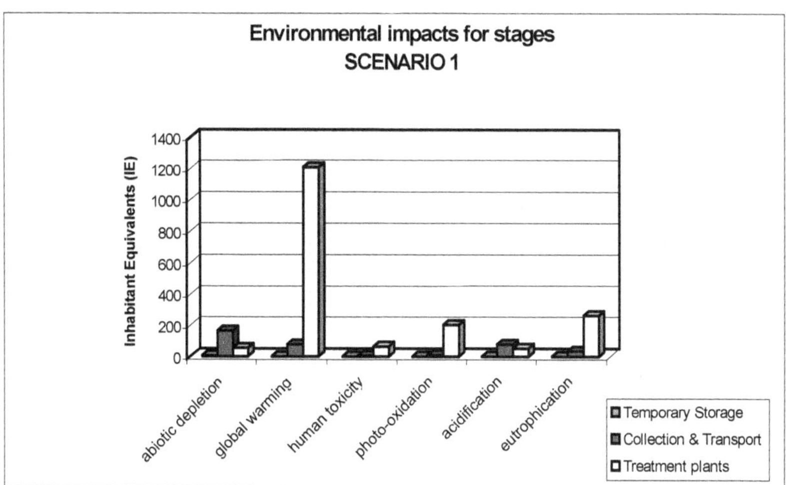

Figure 82. Environmental impacts for subsystems – Reus Scenario 1

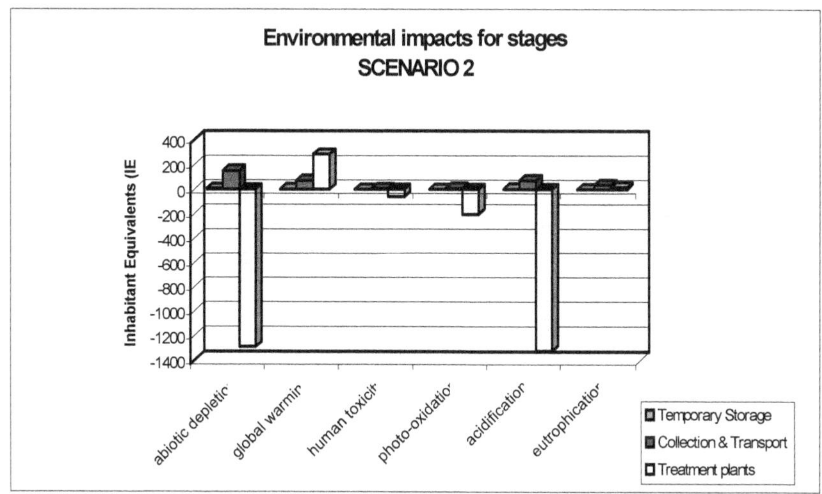

Figure 83. Environmental impacts for subsystems – Reus Scenario 2

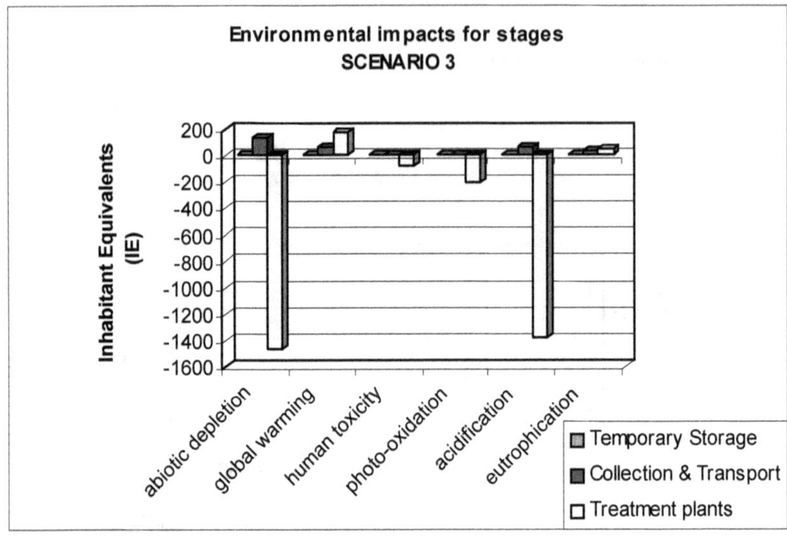

Figure 84. Environmental impacts for subsystems – Reus Scenario 3

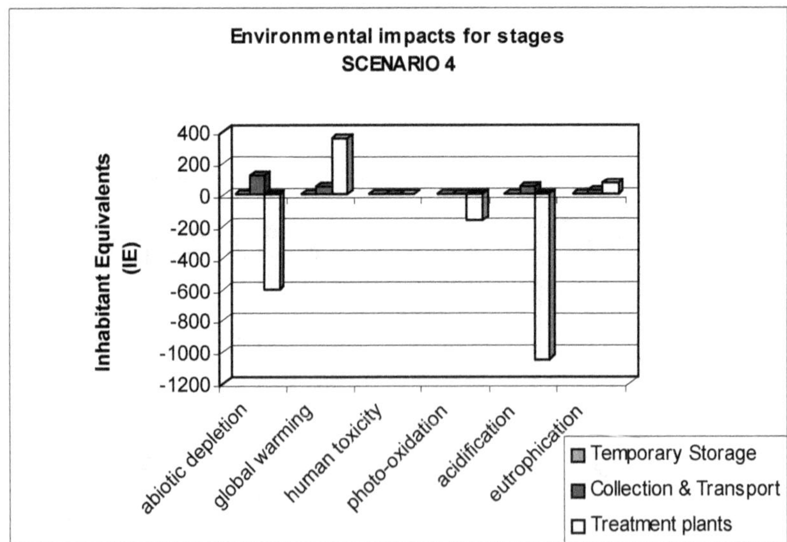

Figure 85. Environmental impacts for subsystems – Reus Scenario 4

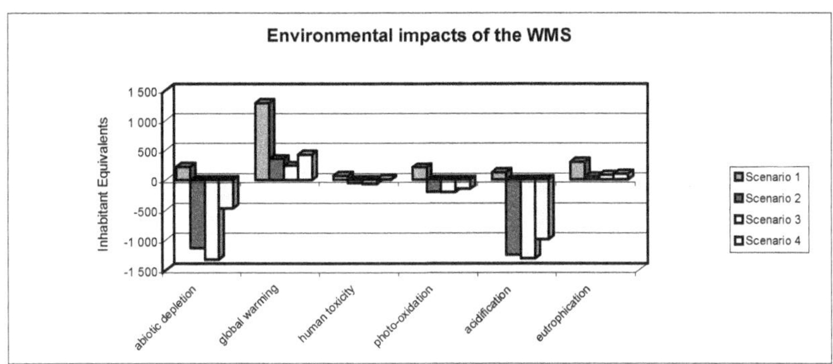

Figure 86. Environmental impacts of MSWMS in Reus

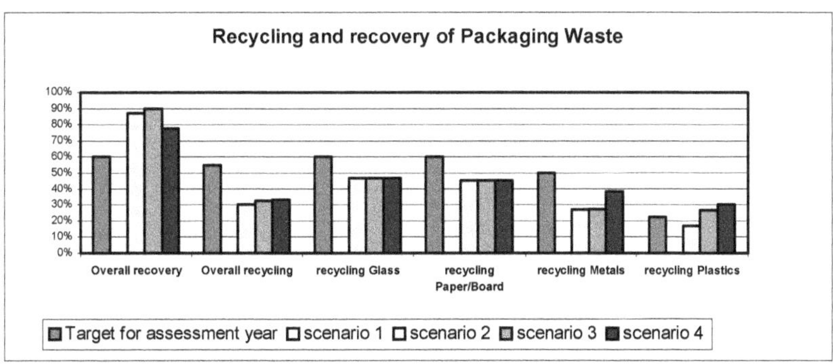

Figure 87. Recycling and recovery of Packaging Waste for Reus

Economic Assessment

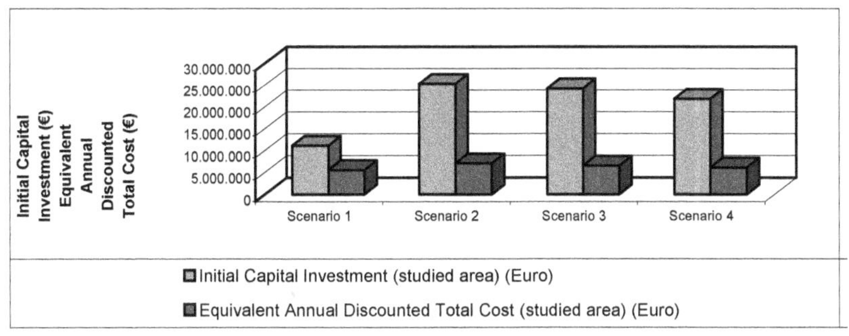

Figure 88. Economic impacts of MSWMS for Reus

Table 49. Economic Impacts of MSWMS in Reus

Scenario 1	Scenario 2	Scenario 3	Scenario 4
Initial Capital Investment (10^6 €)			
11,1	25,3	24,2	21,8
Equivalent Annual Capital Cost (10^6 €)			
5,6	7,1	6,6	6,1

Table 50. Economic Efficiency at the Municipal level (Reus)

	Scenario 1	Scenario 2	Scenario 3	Scenario 4
Cost per ton (€/ton)	95,4	120,69	111,81	111,86
Cost per household (€/hh)	131,6	166,50	154,26	143,53
Cost per person (€/person)	57,19	72,37	67,05	62,39
Revenue from recovered material and energy (€)	0	508.484	488.935	1.005.820
Total cost as % of GNP of the city (%)	0,42%	0,53%	0,49%	0,46%
Diversion between revenue and expenditures for MSWMS (%)	33,22%	115,10%	124,23%	133,52%

Table 51. Economic Efficiency at the subsystem level (Reus)

Subsystem	Cost per ton of waste (€/ton)			
	Scenario 1	Scenario 2	Scenario 3	Scenario 4
Temporary Storage	3,00	3,00	2,77	2,81
Coll. & Transport	67	70,76	58,75	59,93
Landfill	25	0	0	0
Composting	0	0	49	48
Digestion	0	0	0	0
MBP aerobic	0	0	0	0
MBP anaerobic	0	0	0	15
Incineration	0	31	33	38
Paper sorting	0	81	81	81
Glass sorting	0	189	189	189
Metals sorting	0	0	0	0
Plastics & comp. Sort.	0	0	0	0
MDR sorting	0	93	93	93
WEEE sorting	0	141	141	141

Table 52. Equity and Dependence on subsidies (Reus)

	Scenario 1	Scenario 2	Scenario 3	Scenario 4
Cost per person as % of minimum wage	334,42%	423,23%	392,11%	364,84%
Cost per person / income per person	0,39%	0,50%	0,46%	0,43%
Subsides or grants per person (€/person)	32,64	37,49	34,95	32,86

Social Assessment

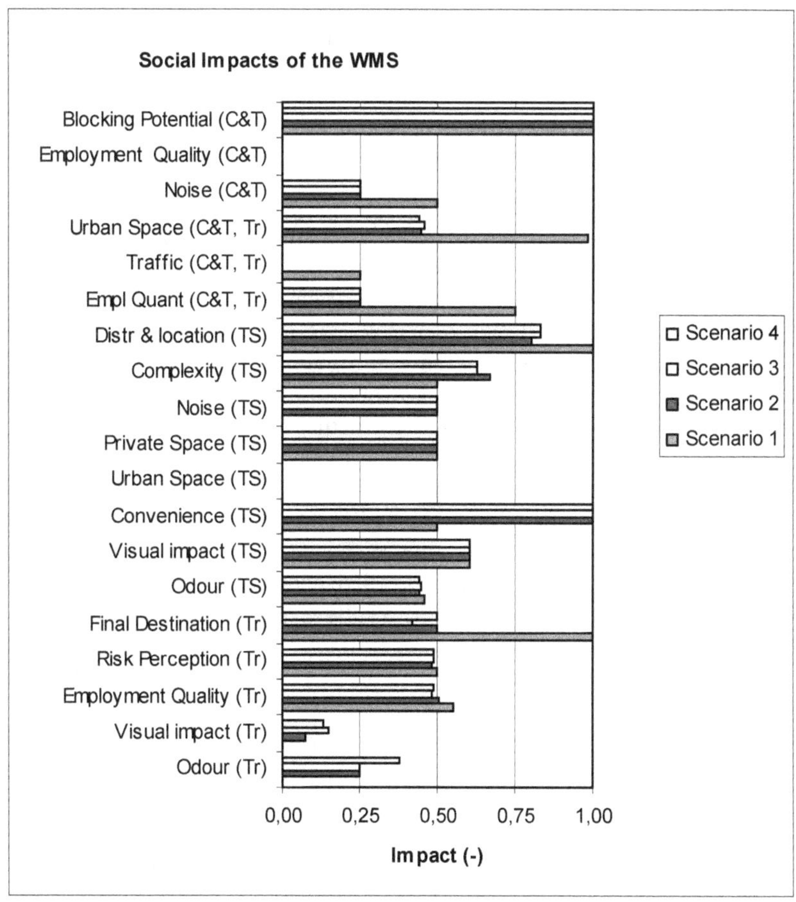

Figure 89. Social Impacts of the MSWMS in Reus

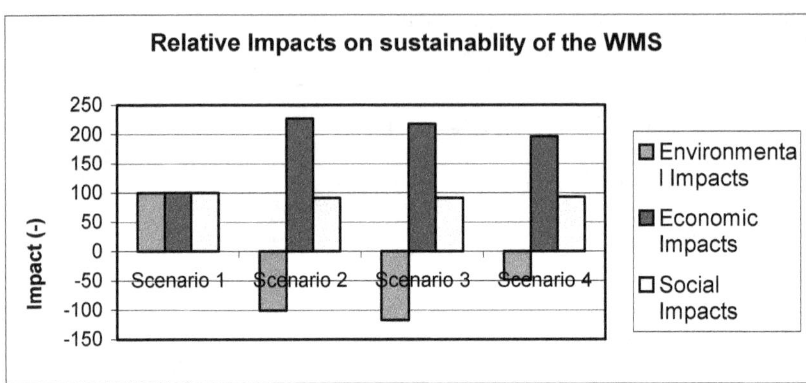

Figure 90. Relative Impacts on sustainability of the MSWMS for Reus

Conclusions and recommendations

In this case study slightly different attempt was used while defining scenarios to be assessed and compared. The reason for that was actual waste management strategy of the municipality of Reus (Scenario 3) which, unlike opposition to other cities, is quite advanced and well implemented. Thus it was decided to define two scenarios which would represent past situations and another one looking into a hypothetic future. The two scenarios reflecting past situations were very different one to another and the fourth one was not much different to the actual one.

As in all modelling cases, results should be looked at carefully, because they will be as accurate as input data of the model and also because they will be more realistic the closer the model is to reality. In this case, the amount and variety of input data needed coupled with a very complex model, make the interpretation of the results a very tough task.

Having said that and looking into the previous graphs it can be concluded that:

- Scenario 1 (only landfilling of untreated waste) shows a very clear different behavior with regard to environmental, social and economic impacts, since the definition of this scenario is clearly different to the others. This concept of waste management entails environmental burden (environmental impact is positive) while the economic impact is lower than that of the other scenarios (relatively low cost of simple landfilling of waste),

- Scenarios 2 and 3 show very similar values for their environmental, economic and social indicators but scenario 3 (actual situation of waste management) is slightly more advantegous taking into account all impact categories,
- Further development of waste management from scenario 3 to scenario 4 entails improvement of sustainability of waste management with regard to environmental and economic impact categories but this concept could be socially less accepted.

None of the scenarios assessed ensure achieving desired recycling rates of packaging waste – this remark concerns each material flow as well as overall recycling of packaging waste. Overall recovery rate of packaging waste have been achieved in scenarios 2-4 due to energy recovery during incineration of residual municipal waste.

When using the tool to decide on waste management strategies, rather than looking to the overall environmental, economic and social impacts, municipalities should previously define which of the indicators are more important. Are they more related to their Local Agenda 21 or to a some sort of internal policy implemented.

10.3 Summary of results of case studies

Four to five alternative waste management scenarios, reflecting various situations and predictions of the waste management were developed for each of five European cities in close collaboration with the local municipality representatives. One scenario (generally No. 1 but in Reus No. 3) characterises the present situation, while the other three or four analyse some possible developments (characterised by different solutions of Temporary Storage, Collection, Transport and Treatment). Obviously, there are also many other potential alternatives in each case, not presented here due to limited volume of handbook, which should be assessed using developed tools in order to find the best solution. These could be done during further studies and analyses by the municipalities. Presented results of case studies show the wide range of different potential solutions and demonstrate possibilities of the tools to assess them.

It is a really difficult task to compare among all the scenarios analysed because there are some differences in attempts, quantities of waste managed and many other details. But from the other side all contributors were using the same methodology and Tools developed, which lead generally to a certain unification of case studies. Taking into account the conviction that it is not possible to do a detailed comparison between each case study, an overall outlook of results seems to be interesting and useful for the reader to realise the general differences and similarities between analysed situations.

The present waste management of the target cities, except Reus, is characterised by the relatively low level of recycling and recovery of materials as well as by a lack or small extent of residual waste treatment before landfilling. Reus, with implemented recycling, composting and incineration of waste, presents a relatively high level of waste management development.

Results of case studies done for all target cities have shown that proposed modernisation of existing waste management systems will generally entail environmental relief and improving social acceptance of future waste management scenarios in comparison to the present situation. This conclusion is not fully valid for Reus where adding of MBP to existing system will result in a slight decrease in social acceptability of the proposed waste management system. However, improvement of waste management system needs investments in waste treatment facilities and increased costs of operation for these facilities.

Generally, investment and operation costs increase with growing degree of waste treatment before landfilling; incineration generally entails higher costs than aerobic mechanical-biological pretreatment of waste.

In each study the most simple waste management system (landfilling of untreated waste as main disposal option) is portrayed in reference scenario 1 (in all cities except Reus this is the present system), while the other scenarios contain different solutions of waste collection, transport and treatment. Generally, the extent of separate collection of materials, levels of recovery as well as degrees of waste transformation before landfilling increase with growing scenario number (from 2 to 4 or 5 (Xanthi)).

As for the general conclusions concerning environmental impacts, most scenarios of waste management, including actual ones (except Reus and Xanthi), show global environmental advantages due to the credits allocated to recovery of materials and energy (negative impacts). Generally (with small exceptions), with growing scenario number its environmental relief increases. Treatment stage of waste management is contributing mainly to this effect.

It should be noted, that aggregated environmental impacts are the sum of all impacts in IE for all considered categories, without applying weighting factors or differentiating the importance of the environmental indicators.

The most important environmental burden (positive impact) of the existing waste management systems in target cities (except Reus) is caused by emissions of greenhouse gases (mainly methane from landfills) which is expressed by the impact category "global warming". The other stages, calculated together, show much smaller environmental impacts in all impact categories, except the one above mentioned. The impact categories abiotic depletion and acidification in the treatment stage present much bigger negative impacts (this means reliefs) than positive impact of global warming, thus the condensed environmental impact is negative. In the case of Reus, simple landfilling of untreated waste (scenario 1 reflecting past situation of waste management) leads to environmental burden with regard to all impact categories.

Analysed scenarios of the future waste management show in most cases high negative impacts in categories: abiotic depletion, acidification, global warming and photo-oxidant formation while impacts in the categories: human toxicity and eutrophication are much smaller (negative or positive, depending on scenario). The aggregated environmental impacts of prevailing number of scenarios are highly negative – this means environmental relief. Table 53 contains aggregated environmental impacts of each scenario, expressed in IE.

Table 53 Aggregated Environmental Impacts for all scenarios

City	Popul. inh.	Scenario 1	Scenario 2	Scenario 3	Scenario 4	Scenario 5
Kaunas	374000	-4009	-11205	-13304	-12114	
Nitra	87000	-509	-1167	-1496	-1736	
Reus	98000	2218	-2231	-2591	-1082	
Wrocław	640000	-6669	-26959	-31367	-31673	
Xanthi	45000	635	-2197	-30,6	-3447	-4121

Specific value of aggregated environmetal impact related to one citizen of each municipality amounts from -0,02 to -0,11 IE.

Regarding social impact assessment it should be noted that this matter is very sensitive and needs a lot of additional studies, concerning particular criteria and indicators, which should be conducted in the cities to obtain reliable answers. Results of preliminary studies, presented in this chapter, should be verified during following surveys of public opinion, repeated at regular intervals, in order to assess the changes in social acceptance as a result of public education. Preliminary assessment of social impact of analysed waste management scenarios showed in most cases increasing acceptance of proposed scenarios of waste management modernisation in comparison to scenario 1.

Table 54 presents social impacts of each scenario in the analysed cities confirming the results of preliminary assessment. But it is still difficult to clearly interpret the results obtained between the cities and also between the scenarios assessed within the city. While analysing the results of social studies on the level of municipality the "hot spots" should be identified and taken into account when preparing the program of social education.

Table 54. Aggregated Social Impacts for all scenarios

City	Popul. inh.	Scenario 1	Scenario 2	Scenario 3	Scenario 4	Scenario 5
Kaunas	374000	9,9	10,2	9,8	9,7	
Nitra	87000	8,5	7,2	8,0	7,2	
Reus	98000	9,1	8,3	8,3	8,4	
Wrocław	640000	11,0	9,5	10,2	8,3	
Xanthi	45000	8,7	7,6	7,9	7,9	7,2

Economic assessment of waste management scenarios show similar general patterns and relationships within the cities, and generally also between them, but the cost levels are different, according to different economic conditions of each EU country, some differences betweeen analysed scenarios and local conditions. In each city the initial capital investment cost as well as equivalent annual discounted cost were the lowest for reference scenario (main treatment option is landfilling of untreated waste), while the scenarios containing incineration of waste entail, in most cases, the highest costs. Detailed economic data for each scenario assessed are presented in particular subchapters but Table 55 gives the comparison of ranges of some economic indicators for analysed waste management scenarios for target cities.

Table 55. Ranges of some Economic indicators for the considered cities

City	Popul. inh.	ICI, mln €	EADTC, mln €/year	Cost per ton €/ton	Cost per person, €/person	Cost as % min. wage, %	Cost as % of ind. income, %
Kaunas	374000	30-50,2	8,6-14,3	56-94	23-38	371-630	0,54-0,92
Nitra	87000	12,3-16,4	4,1-4,9	99-119	47-57	473-568	1,4-1,7
Reus	98000	11,1-25,3	5,6-7,1	95-121	57-72	334-423	0,39-0,50
Wrocł.	640000	44-116	12-22	39-71	19-34	196-362	0,27-0,50
Xanthi	45000	2,7-12,0	0,8-2,2	52-140	18-47	60-157	0,08-0,22

This comparison shows some typical ranges and trends of costs which reflect economy of scale of capital and annual cost of waste management in the target cities, even taking into account earlier mentioned differences and limitations of such comparison.

Assessed scenarios of municipal waste management in Xanthi, the smallest municipality, are characterised by relatively hight unit costs per ton of waste treated but simultaneously also by the lowest cost per citizens in relation to minimum (60-157 % of daily minimum wage) and average (0,08-0,22 % of average annual income) income. By comparision Nitra has the highest cost per citizen (473-568 % of minimum daily wage and 1,4-1,7 % of average annual income).

It is visible that relative differences between unit costs of waste management within similar solutions (for example within reference scenario on one side or within the scenarios with thermal waste treatment on the other side) are much smaller than relative differences between indicators showing waste management costs as percentage of individual incomes (average or maximum). The reason for such big variety of values of economic indicators characterising equity of waste management systems assessed result from the differences in individual incomes between citizens in the EU 15 countries (Greece and Spain) and those of the new members of the EU (Lithuania, Poland and Slovakia).

11 Good Practice in Planning of Waste Management Systems

11.1 Introduction

The developed Tool assesses the sustainability of waste management systems considering environmental, economical, and social indicators. However, there are many other important aspects, which cannot be measured or are difficult to measure by any indicator. In this part of the handbook, these aspects have been discussed and example cases from practices have been given.

11.2 Management capability

Although the basic principles of waste management remain unchanged, from day to day new technologies are developed for the implementation of these principles. These developed technologies become more complicated over and again and they should be used in accordance with changed legislations, financial, economical and social aspects. This needs a capability of management to afford making right decisions at the right time.

Figure 91 shows the different management assignments within a modern waste management system. The duties to be done vary very strongly and need skilled personnel in each stage of the management.

Technical management is necessary for the operation of the whole system, particularly treatment and disposal processes.

Within the financial management, the account of all revenues and expenditures is kept. Financial manager is responsible for the procurement of equipment, replacement part and other operational material. The taxes for waste management and payments for extra services are controlled. Revenues from the selling of recyclables are recorded. Furthermore, he should prepare quarterly and annual reports and balances.

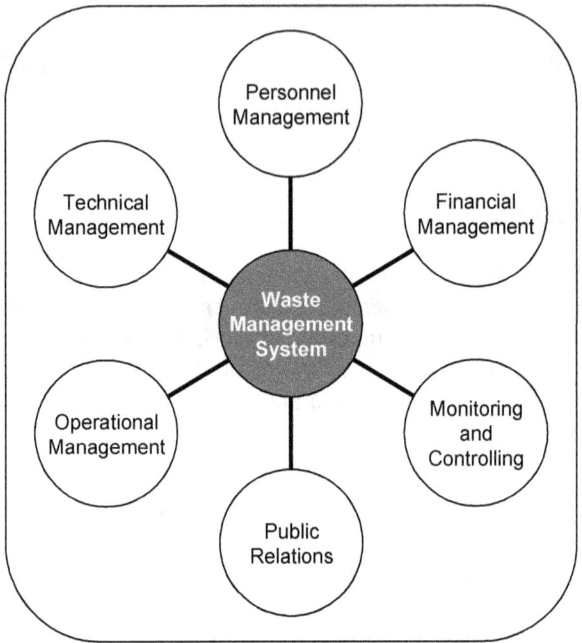

Figure 91 Management assignments within a waste management system

The tasks of operational management are the organisation of the collections routes, time schedules and maintenance and repair of the equipment. In addition to household wastes, the collection of clinical, hazardous and other specific wastes should also be ensured. The concept for the collection of recyclables could be further developed according to local and temporary circumstances.

Personnel management is responsible for the administration of personnel affairs, which includes management of wages and salaries, taxes and social security contributions, recruiting of new personnel and their training. Furthermore, the preparation of staffing plans and job descriptions is another task of personnel management.

For a successful operation of the waste management system, the waste quantities and compositions should be analysed and forecasts should be made at an early stage. Furthermore, the processes of waste treatment and disposal should be monitored and the emissions should be measured continuously. All these tasks require a monitoring and controlling system in addition to be constructed facilities and purchased equipment. While the waste quantities could be determined due to weighing all waste deliveries to the treatment facilities, the waste composition could be analysed every two years.

Public relations activities are one of the important tasks within a modern waste management system. Implementation of a new system can only be successful if the public accept the system and take part in the system. Therefore, the department of public relations should develop information materials, organise events, school visits, etc. For the presentation of the information material, local media, newspapers, TV's and radios can be used. A further task of the public relations is the gathering of the complaints and desires of the public and analysing the aspects as willingness to pay and affordability. Qualified staff, which has pedagogical abilities and knowledge of waste management issues is required in the implementation of public relations. Additionally, experiences in dealing with public media are of major importance.

Since these tasks are very complicated, the personnel should be skilled and experienced. Therefore, it is the duty of the decision makers to check the availability of the management capability before they decide in favour of a new waste management system.

For all above listed tasks, different persons from different branches are needed: managers, engineers, economists, sociologists, technicians, drivers for specific vehicles etc. In addition to their skills and education level, the gained experiences with waste management are also very important. In this respect, availability of a university with waste management department and laboratories is profitable for the waste management system.

The complexity of waste treatment processes is shown in Table 56 taking the mechanical-biological pre-treatment as an example. From the right to the left, the complexity of used processes is decreasing. Compared to anaerobic digestion with its high-level technology, the landfilling of shredded material is a basic technology which can be implemented and operated with relatively low effort. The utilisation of process gas and control of emissions from different stages of the facility would complicate the operation. In addition, a high technology process is difficult to expand since the processes conditions will change with changed quantity and quality of the waste. In this respect, the operation of a high technology needs skilled and experienced personnel.

In addition to the availability of skilled persons, a new institutional structure should be established or the existing structure should be reformed. The appropriate distribution and decentralisation of responsibilities are necessary for a functional waste management. An efficient coordination within the waste management department and with other departments of the municipality should also be established.

Table 56 Comparison of different mechanical-biological pre-treatment processes (IKW and I&U 2000)

Process	Units	Anaerobic digestion	Intensive rotting	Biodegma process	Natural-draft process	Outdoor heap rotting (Turner)	Shredded-refuse landfill
Mechanical conditioning		screening, comminution	screening, comminution	screening, comminution	comminution (screening)	comminution (screening)	comminution only if necessary
Heap / digester							
Duration (encapsulated)	weeks	3	12 - 16	4	-	-	-
Duration (unencapsulated)	weeks	2 - 6	-	8 – 12	24 - 32	24	> 32
Aeration		passive (post-rotting)	active	Active	passive	by turning	diffusion on the surface
Exhaust collection / cleansing		biogas utilisation	Biofilter or thermal	Biofilter or thermal	heaps covered with biofilter material	-	-
Costs (German prices)							
Initial investment	Mio. €	15 - 20	10 - 15	4 – 6	1 - 2	0.5 – 2	0.3 – 0.6
Initial specific investment	€/ ton,y	300 - 400	250 - 350	100 - 150	15 – 40	20 – 45	5 - 10
Treatment (spec.)	€/ton	60 - 80	55 - 70	30 - 45	5 - 25	8 - 30	5 – 15
Space requirement	m^2/ton*y	0.2 – 0.4	0.4 – 0.8	0.4 – 0.5 (without post rotting)	0.6 – 1.2	0.5-1	2
Energy requirement		autarkical	high	medium	low	medium	low
Odour emissions		during post-rotting, turning (biofilter)	biofilter	during post-rotting, turning (biofilter)	construction, reduced emissions in the rotting process	construction, turning, entire rotting process	construction, entire rotting process
Level of technology		very high	high	medium	low	low	none
Maintenance req.		very high	high	medium	low	low	none
Incremental extensibility		high effort	high effort	good	good	good	good
Flexibility		low	low	high	high	high	high

11.3 Public awareness

The acceptance and co-operation of the public is very important for the successful implementation and operation of a new waste management system. However it is not possible to predict the exact behaviour of the public before a new waste management is established and it is also difficult to find out the shift when something in the system is changed. In this account, the planned waste management system should consider the existing structures of the social life of the public and where necessary optimise them.

The legal measures and penalties may change the public's behaviour and obligate it to comply with regulations of new waste management system. However, causing public awareness and providing information may be the first step to an increasing acceptability of the new system by the public. At the beginning of a project, public awareness activities should immediately be started to prepare the citizens for altered conditions of waste management.

Planned measures for a better waste management are mostly linked with additional financial burden for the waste generators. To achieve the acceptance of the new waste management system, the following conditions should be fulfilled:

- an efficient and reliable waste collection and disposal, which fulfils the wishes of different groups of the public,
- an optimised payment structure: just distribution of costs according to polluter pays principle and higher taxes for wealthy population groups according to solidarity principle,
- realizing of the planned long-term macroeconomic benefits

It is also very important, that the public recognise the need for waste treatment and disposal facilities of all kind. The technical standards, as minimal distance to the dwellings or measures to minimise emissions should be observed, so that the public is not interfered by waste treatment.

Trying to develop the awareness of the population and the waste producers, it is important to take respective needs and sensitiveness of the social stratums into account. Private households can be differentiated e.g. by:

- income class
- age group
- gender
- education level
- settlement structure

For other waste producers special public relation activities can be attempted. If necessary, e.g. at hospitals, individual and subject based consultations should be implemented.

Examples for the impact of age and gender regarding recycling attendance can be given from a study in England. 50% of people aged 35 and over say they recycle more than 40% of their household waste. In contrast, only 35% of people under 35 separate this amount of waste. In households, where recycling takes place, 69% of women have the responsibility compared to 48% of men. While a quarter of men share the responsibility with other people in their households, only 6% of women say their spouses take responsibility (MORI 1999).

The education level of the public is also an important factor to understand the waste management system or recycling respectively. The higher the education level in a household so is the higher rate of recycling in this household (Jenkins 1999). The findings of another study from England show that the parents with children of school age receive information through their children about recycling and waste. Two thirds of this group – 10% of adult population in England – think that it has influenced their household and behaviour (Waste Watch 1999).

The reasons for recycling vary per each person. In Table 57 the findings of the MORI-study in respect of this topic are shown. Around 94% of asked 2,005 persons said that the recycling is worthwhile. However, they gave different reasons for their opinion. Around 86% thought that recycling helps protect the environment and saves resources. Around a third said recycling reduces the amount of waste going into landfill.

Table 57 Reasons for the recycling (MORI 1999)

We recycle because	Percentage [%]
It is our way of helping the environment	54
It is what every responsible citizen should do	32
It is just a habit	26
It is so easy it would be stupid not to	16
Our children make us	5
None of these	3
Not stated	8

There are various reasons for not participating in recycling. The most given reasons are too lazy or too busy. In the second range stands the lack of facilities, distances to the container or inconvenience of the collection system.

The following subjects and actions can be defined with regard to public relations:

- information about altered collection system and recycling: new containers, collection vehicles, time schedule, separate collection, etc.
- information about new treatment and disposal plants and their necessity: recycling facility, composting plant, landfill
- counselling about and making people more sensitive to waste avoidance and recycling
- motivating to participation
- information about the costs of waste management and explanation of the resulting necessity to increase taxes or raise the fees

Public relation activities can be carried out with different instruments shown in Figure 92.

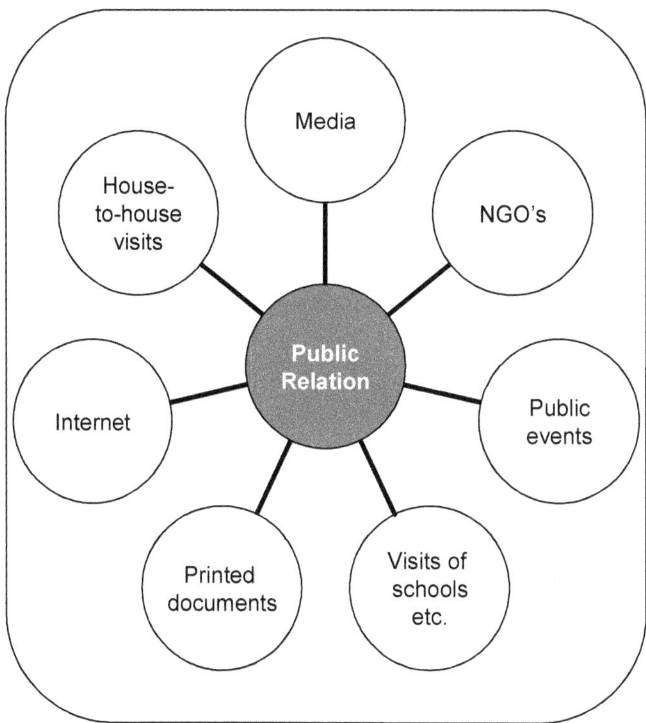

Figure 92 Instruments of public relation

Local and regional media can be utilised as communication means. In television and radio, small spots can be broadcasted. These should be dependent on the respective target group and should convey the aims of the waste management concept as well as necessity of participation of the public. Newspapers should regularly report on the proceedings of implementation of the new waste management concept as well as the advantages which arise for the public and environment. In small cities, public awareness campaigns can be undertaken through house-to-house visits.

Additionally, different groups and organisations, especially the environmental Non Governmental Organisations (NGOs) could be involved. These should be informed regularly and intensively, so that they can spread the knowledge to the public. Furthermore, common events which refer to waste management can be organised. Examples are district festivities, soccer games, concerts, competitions, open day of the waste management facilities, guided tours on the landfill site etc.

Moreover, posters, leaflets and easily comprehensible brochures, which include general and specific questions and recommendations for the respective target groups, form the basis of successful public relations. Advertising on collection vehicles is another way to spread the information. A calendar, which includes dates of collection and some other useful information about waste collection, can be distributed to the households. An internet site with useful information about new waste management system: waste management plan, time schedule for waste collection, benefits of recycling, list of recyclables etc., can also be created. An interactive page on this site can give the users the opportunity to ask questions and make suggestions. For this purpose, a hotline telephone can also be installed. The answers of the frequently asked questions could be published on the internet site.

11.4 Scavengers

Since the waste management systems in developing and transition countries are not well developed, the recycling is mainly carried out by informal sector. Thus, it has another relevance compared to developed countries. Whilst in developed countries recycling represents willingness to protect the environment and it is promoted with economical benefits, in developing countries it is an economical necessity for many people.

Scavengers are mostly unregistered persons from ethnic minorities or poor population groups, who collect the waste either on the street or on the dumpsite. The structure of scavenger groups varies from city to city.

Figure 93 Scavenger on the dump site of Wroclaw (2000).

They work as individuals or families. In some cases, they have unofficial organisations. While one group picks up the recyclables on the street, another group collects on the dumpsite. In some cities each district is served by another scavenger group comparable to normal trading areas.

They collect the valuable material, which they can transport and sell easily. For example, scavengers collect plastic, metals and cardboards but they do not collect organic wastes. They do not have any measurement to protect their health nor have any social or health insurance. They mostly live closely to the dumpsite, since the recyclables should be picked up before next delivery comes and the waste is covered. In this respect, they have mostly long working hours and sometimes during the night depending on waste collection schedule.

Since these people do not have any other job nor even an education, they should be integrated into the new system. An exclusion of these people from the waste management system means more poverty and risk of their existence respectively. The integration of informal sector to the formal waste management has following benefits:

- safeguard employment and regular income for scavengers
- improvement of health and social security conditions for scavengers

- assure the acceptance of scavengers by the public
- supporting communities by recycling activities
- systematising the collection and increasing the recycling rates

In scavenger families, the children undertake the same work as adults and do not go to school. Contracts with the authority can assure the family income and may result in a better education for the affected children.

To these benefits, the general benefits of recycling can also be added. Due to recycling, the lifetime of waste disposal facilities can be expanded, environmental damages are reduced and the raw material resources are conserved. Poor people can improve their living standards due to purchasing of products made from recycled materials for lower prices.

11.5 Other aspects (demands on Temporary Storage)

To get tailor-made solutions within waste management, an efficient investigation of local circumstances is essential.

A waste prognosis model considers principally historical development of waste quantity and quality and not seasonal fluctuations. In some cases, the prognosis and the waste management plan should be adjusted according to local circumstances. For example a city with a lot of tourism might have more waste and probably with other composition in the busy season than in the normal season. Because of increasing waste generation, the waste should be collected more often in the busy season. Since the quantity of biowaste due to high consumption in restaurants will also increase, the capacity of the composting plant should be increased too. In other cases, due to commuter from neighbouring communities, a commercial centre might have more people in daytime than at night. If the waste quantity is forecasted, the population which generates the waste should be considered instead of the population which was counted in the last census.

A necessary fact to be considered is the economical condition in the project region. Before a decision is made on new SWM, the affordability and willingness of the public to pay should be analysed. If there are sufficient financial resources, all suitable techniques can be applied. However, in a region where the citizens and waste management representative respectively do not have the economical affordability for high-tech solutions, simply, man-powered solutions can be used. Using hand sorting instead of expensive sorting machines or turning of compost windrows per hand instead of high-performance windrow turners will decrease the investment and operation costs and thus the disposal costs per waste generator. Furthermore, due to such a planning, additional jobs can be created. These solutions can particularly be considered in rural areas where the waste generation is not very high.

Another important aspect is the understanding of the Temporary Storage system. As explained in Section "Public Awareness", a successful implementation of a recycling system depends on a certain degree of public awareness. In this respect, the Temporary Storage at the beginning phase should not be more complex than the public can understand and accept it.

As a part of recycling, different containers are provided for the collection of different materials. The public has to know the waste types and be able to differentiate the containers. They should know what container volume they need and their living area should have enough places for different bins. The site of container and collection points should also be known to the public.

In regions of public economic weakness, the theft of containers can be a problem. If the containers are provided for a fee, they may be stolen by the public, which is not able to pay this fee. Another case is if the containers are very attractive they can be stolen for other domestic uses. Therefore, the material and appearance of the containers should be chosen depending on local circumstances.

There are criteria for the selection of containers, which would change the behaviour of the public and motivate it to participate in waste separation. For one fraction of the waste uniform containers should be provided, however these containers can easily be differentiated from other containers due to shape, colour or labelling. The heating system in the serviced area should also be considered since plastic containers cannot be provided if in the region ashes are produced. The containers should have a lid which can be simply operated. Big and heavy containers should have wheels and container locations should be paved.

The local authorities or waste management companies should ensure that the containers are always in good condition. A broken lid can discourage the public to put the waste into the containers, thus the waste is mostly put near the containers. Therefore the local market should select the damaged containers for repair and hasty replacement.

Another important aspect is the location of containers. The waste generators should know where the different containers for recyclables are and reach them easily. The visibility of containers should be supported by signs on main streets. The locations near shopping centres or on the streets to the shopping centres are good possibilities to encourage the public to bring the recyclables to the recycling container. A good way is the establishment of collection points where several recyclable fractions are collected. However, for this option there has to be enough space on public ground to ensure good accessibility for the people and collecting vehicles. Another point to be considered is the distance of the collection point to the dwellings. If containers and collection points are only reachable by vehicle this would decrease the accessibility of the public which does not have a vehicle.

Containers on the sidewalks should not disturb the public. Narrow streets are not compatible for big container or collection points. Since there should also be a sufficient volume of container available, the collection frequency should be adapted and small containers should be emptied more often.

Waste collection trucks should have enough places on the street, if they stop to empty the container. The workers should be able to move the container to the truck without great effort. It should be also possible that they clean the container's stand after emptying. The cleanliness of the stand is also an important factor, because in this case, if the stand is not clean, the public will reject the stand and want it to be changed.

12 Waste Prognostic Tool guide

12.1 Introduction

The high variation of waste generation rates as well as growth rates and the lack of knowledge about the impacts on them, usually complicate the estimation of future waste generation rates. Thus over- or undercapacities of facilities often lead to extra costs for the waste management infrastructure. Higher reliability of future estimations can help to reduce these additional costs. Additionally future impacts on the environmental and social conditions in a region can be assessed in a better way.

The LCA-IWM Waste Prognostic Tool is an estimation tool for the future generation of municipal solid waste in European cities. Thereby the focus lies on cities in rapidly growing economies, such as southern and eastern European countries. It enables substantiated forecasts of waste generation rates as well as waste composition estimates. Thereby the consideration of long-term changes of the demographic, social and economic border conditions of a region permit significantly higher forecasting accuracy than usual estimates. Within the framework of this project a prognosis horizon of 10 years was assumed.

In comparison to the usually applied simple methods, such as trend extrapolations, this tool features the following benefits:

- **Consideration of regional and national indicators**: Regional peculiarities of the demographic, social and economic conditions are therefore taken into account.
- **Provision of default values**: The user will be supported in the case of missing data about its region. This enables a trade off between accuracy and practicability for the user.
- **Validation for the broad sample of European cities**: The underlying model was tested for 55 of 90 major cities in Europe. Thus high reliability can be guaranteed.

12.2 Installation

12.2.1 System requirements

The following minimum requirements have to be satisfied:

- Operating platforms
  - Microsoft Windows
    - Windows 98
    - Windows ME
    - Windows NT 4.0
    - Windows XP Home
    - Windows XP Professional,
    - Windows 2000 Professional
    - Windows Server 2003
  - Linux
  - Solaris OS (Solaris 7, 8, or 9)
- Processor requirements
  - Pentium 166 MHz
  - 32 MB physical RAM
- Free hard-disk space
  - Application: 3 MB
  - Java Runtime Environment (v 1.4.2): 80 MB
- Software (Installation see below)
  Java Runtime Environment (JRE), Version 1.2 or newer

12.2.2 Downloading the tool

The LCA-IWM waste prognostic can be downloaded from www.lca-iwm.net

The downloaded zip-file has to be extracted and saved on the user's hard-disk. The folder MswPrognosis_LCA-IWM contains the following:

- 'MswPrognosis.jar' contains the Java-based application. In Windows-based operating systems, the application can be started with double-clicking on the filename. Otherwise it can be started using commands in DOS mode (Windows), or in Shell scripts (Linux).

- "Scenarios folder" contains datasets which can be loaded by the user. It contains completed exemplary scenarios.

- 'ReadMe.txt' contains information about the installation of the tool as well as the Java Runtime Environment.

- The other folders ('Images' and 'net') contain resources for the application itself. They are hidden in Windows systems and should not be removed, deleted or altered.

12.2.3 Downloading Java Runtime Environment (JRE)

To run the application, it is necessary to have Java Runtime Environment installed on the hard-disk of the computer. As it is an important Web-based program, it is very often, but not always, available on customary personal computers.

What is Java?

Java is a widely used programming language which offers the following advantages for the user of a Java application or applet:

- **Independence of the operating platform**: It is usable under Windows, Linux, Unix and others, independent of the settings of the operating system.
- **Full functionality**: No additional programs are necessary.
- **High robustness**: Low probability of abnormal ends of the program.
- **Free download of the program via Internet**: The download is free of costs and may be used legally even for commercial use. More information about the license agreements are available, e.g. at http://java.sun.com/j2se/1.4.2/download.htm.

In the following the installation for Windows and Linux is described. It should be taken into consideration that the layout of the application is optimised for Windows systems.

Installing JRE for Windows

Firstly one has to download the Java Runtime Environment. The actual JRE version 1.4.2 is available for free at

http://java.sun.com/j2se/1.4.2/download.html
(Go to J2SE v 1.4.2_06 JRE)

Now one has to execute the resulting setup-program. It is recommended to edit the environment-variable PATH afterwards, but this is not mandatory. If one changes the environment variable PATH one has to edit the "AUTOEXEC.BAT" or to modify the environment variables settings in one's system control. PATH has to point to the executable program files and to the Java-libraries. So the new entry could look like this:

```
SET PATH=C:\Programme\Java\j2re1.4.2\bin;
C:\Programme\Java\j2re1.4.2\lib;%PATH%
```

or

```
SET PATH=C:\Program Files\Java\j2re1.4.2\bin;
C:\Program Files\Java\j2re1.4.2\lib;%PATH%
```

Here C:\Program Files\Java\j2re1.4.2\ is the position where the setup-program installed the complete JRE. Additional installation information is available at http://java.sun.com/j2se/1.4.2/jre/install-windows.html.

Installing JRE for Linux

Firstly one has to download the Java Runtime Environment. The actual JRE version 1.4.x is available for free at
http://java.sun.com/j2se/1.4.2/download.html.

If one has not downloaded the RedHat RPM one has to unpack the archive and copy the extracted directories to /usr. How this can be done for the different available Linux-distributions can also be read at
http://java.sun.com/j2se/1.4.2/jre/install-linux.html.

From the following some information can be found about the installation when one uses the RedHat RPM. For all other offered packages the procedure is very similar. Typically the JRE was installed in /usr/java/j2re1.4.2. As a first step one has to set the environment variable PATH. This has to be done in "/etc/profile" by adding /usr/java/j2re1.4.2/bin separated by a

":" to the existing PATH-entry. After this step these lines have to look like this:

PATH=$PATH:/bin:/...:/usr/java/j2re1.4.2/bin/...
export PATH

Now one has to set the environment variable PATH to the class libraries. This has to be done in "/etc/profile" by adding /usr/java/j2re1.4.2/lib separated by a ":" to the existing PATH-entry. After this step these lines have to look like this:

PATH=$PATH:/bin:/...:/usr/java/j2re1.4.2/bin/:
/usr/java/j2re1.4.2/lib/
...
export PATH

Additional installation information is available at
http://java.sun.com/j2se/1.4.2/download.html.

12.3 Using the LCA-IWM Waste Prognostic Tool

This chapter describes all steps from the program start to the results presentation.

12.3.1 Program start

Program start depends on the operating system:

- Windows only:
  Double-click on the file "MswPrognosis.jar"

- Windows or Linux:
  Use the command

  java -jar MswPrognosis.jar

  as MS-DOS (Windows) or Shell (Linux) command.

Figure 94. Initial screen of the LCA-IWM Waste Prognostic Tool

After pressing a key the user will get to the first input screen. To complete a new scenario, one needs just to fill or modify each field and move from form to form using the 'Next' buttons.

Clicking on the button 'Load a file' enables loading an existing (partly or fully completed) scenario. The loaded dataset can then by modified and saved on each form screen (see Figure 95).

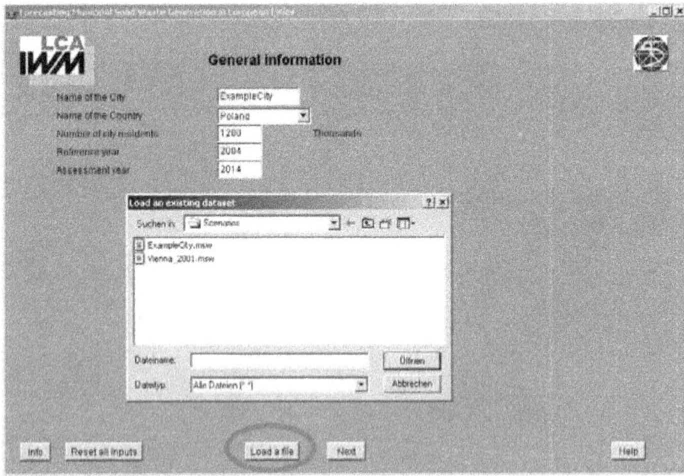

Figure 95. Loading an existing scenario.

12.3.2 Input of the user's data

The user needs to pay attention to the following requirements concerning data inputs:

Each input in a field needs to be confirmed by hitting <ENTER>!
Otherwise the value will not be recognised.

Input values without commas and thousands separator!

Wrong: 1,234.56 Wrong: 1234,56 **Right: 1234.56**

Each shown default value may be overwritten!
If not accepted, the value is out of plausible range (error message appears).

In all cases, helping messages will guide the user through the form screens. Each invalid input will be indicated by red, italic messages.

General information

The necessary general information about the city contains the following data:

- 'Name of the City'
- 'Name of the Country': It is possible to select one out of 32 European countries. The selection has an impact on the offered default values for the urban socio-economic indicators as well as the recommended national indicators (see Chapter 3.2.4 and 3.2.5). If the country in question is not included in the list, it is useful to select a neighbouring country with similar social and economical condition (e.g. San Marino → Italy, or Monaco → France). If this option is not possible, 'Country not listed' on the end of the list has to be selected and no default values will be available.
- 'Number of city residents'
- 'Reference year': Here the last year with available MSW collection data has to be selected. Usually this is the last or second last year from the actual year.
- 'Assessment year': In both LCA-IWM tools, the standard assessment year is defined as the tenth year (Year X+10) after the reference year (Year X). In order to permit other forecast horizons, every year after the reference year up to the year 2020 can be defined in

this tool. It should be considered that accuracy of every forecast decreases with the length of the prognosis horizon.

Municipal solid waste collection quantities

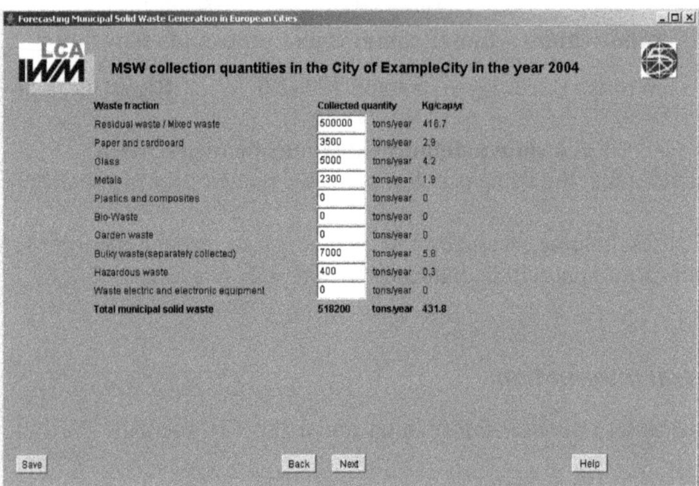

Figure 96. MSW collection quantities.

The municipal waste generation in the reference year marks the starting-point of the waste forecast. The quality and reliability of these waste data can have a remarkable impact on the accuracy of the forecast. The following errors and deviations should be avoided. If possible, these quantitative effects should not be taken into account:

- **A typical short-term impacts**: Floods, stormy seasons or similar events can lead to higher collection quantities of defined waste types (e.g. the impact of floods on bulky waste, or storms on garden waste). As this forecast tool aims to estimate long-term developments, these additional quantities should not be considered. In this case these extra quantities (compared with foregoing years) should be excluded.
- **Major change of number of non-household waste generators**: Extensive systematic changes within the forecasting period, e.g. consideration / non-consideration of collected commercial waste within the total MSW collection, cannot be estimated with this tool,

as the number and type of included waste generators changes. The quantitative basis should be the same (e.g. household waste only, or household waste and wastes from small commerce and official buildings).

- **Wrong allocation at the turn of the year**: Statistical problems leading to the wrong allocation at the beginning and end of the year should be prevented.

The mentioned waste types include the following fractions:

- 'Residual waste / Mixed waste': Commingled waste without separately collected waste. This waste stream includes all other separately collected waste streams which cannot be allocated to an other mentioned waste fraction (e.g. textiles, wood).
- 'Paper and cardboards' (e.g. Newspapers, magazines, letters, writing paper, books, paper bags, cardboard, boxes)
- 'Glass' (Disposable bottles, preserve jars, milk and soft drink bottles, wine and spirit bottles etc.)
- 'Metals' (Food cans, metal foil, metal tubes, tools, cable, drink cans, pipes, steel bands etc.)
- 'Plastics and composites' (Packaging foils, plastic bags, beverage containers, plastic cans and bottles for laundry detergents, containers for shower gels, shampoos etc.)

ATTENTION:

Recyclables (paper and cardboard, glass, metals and plastics and composites) can be collected separately as single waste stream or as packaging waste / Mixed Dry Recyclables (MDR). In each case, it is necessary to collect and input the collected quantity – either the collection quantities (if fractions are collected separately) or as output from the sorting facility (if collected as Packaging waste / MDR).

- 'Bio-waste' (Waste from kitchen or apartment, such as fruit and vegetable waste, peels, leftovers, indoor plants, potting soil etc., and, if not collected separately, also garden waste)
- 'Garden waste' (e.g. cut grass, herbs and weeds, dead leaves, cuttings from trees etc.)
- 'Bulky waste': Commingled waste which is collected in a separate collection schedule due to its bulky form.

ATTENTION:
If all residual waste / mixed waste, including bulky parts of it, is collected together in big sised containers, there is no bulky waste collection in the user's city. In this case enter zero.

- 'Hazardous waste' (A wide spectrum from e.g. cooking oil, medical drugs, mineral oils, batteries to thermometers, cleaning agents, paint or photographic material)
- 'Waste electric and electronic equipment': All devices using electric current.

ATTENTION:
Other waste quantities which cannot be allocated to the mentioned waste types have to be added to the residual / mixed waste!

Residual waste composition

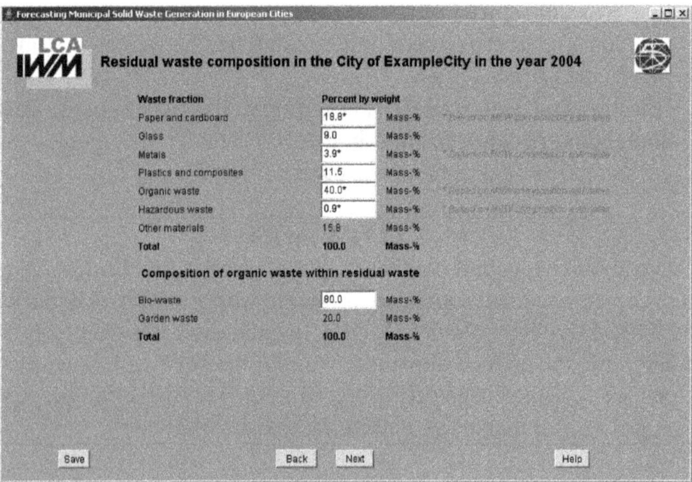

Figure 97. Residual waste composition.

Analyses of residual waste aim to verify the material characteristics of the municipal solid waste which is not collected separately. Simple sorting analyses document at least the main material fractions. The methodology for the procedure is available at the website of 'SWA-Tool', a parallel project within the EU Waste Management Cluster (www.swa-tool.net).

The necessary input data should be based on as much as actual sorting analyses and covers the mass percentage by main fractions:

- paper and cardboard,
- glass,
- metals,
- plastics and composites,
- organic waste, and
- hazardous waste.

If available, the composition of organic waste within the residual waste should be entered.

If available, the input screen shows default values of the residual solid waste composition which can be used in the case of missing data (e.g. concerning hazardous waste). These defaults are based on the estimated total municipal solid waste composition which can be expected under the socio-economic border conditions in the defined country in the reference year. The estimation considers the amount of separately collected quantities of the mentioned fractions (An example: The higher the collection quantity of e.g. waste paper, the lower is the amount of paper in the residual waste).

Actual socio-economic conditions

Municipal solid waste generation and composition strongly depends on the level of affluence. The higher the social and economic status of a city, the higher are the waste generation rates as well as the amounts of waste fractions which are linked with increased consumption, such as paper and cardboard or plastics and composites. Within this project, these relationships have been evaluated quantitatively on the basis on a Europe-wide investigation which covered all major European cities over a period of more than 20 years.

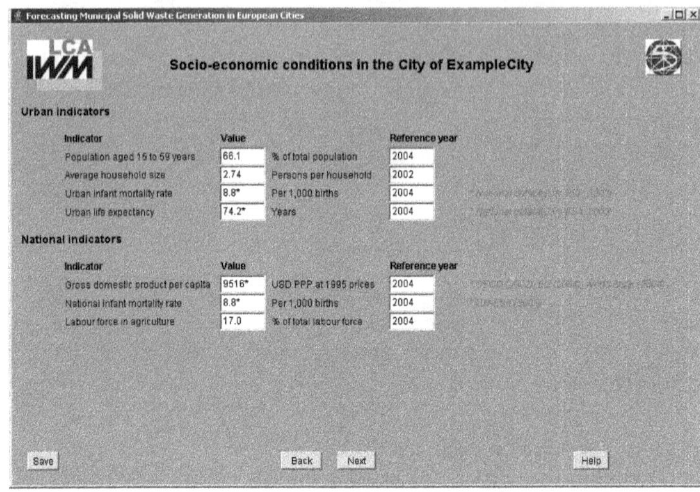

Figure 98. Socio-economic conditions.

Because of this a small set of demographic, social and economic indicators has been used to quantify the level of affluence. The indicators are related to the city itself ('Urban indicators') or to the whole country ('National indicators'). The selection of these indicators is based on good data availability and statistical significance.

Concerning each indicator two figures can be input by the user:

- Value of the indicator: The most actual value should be used as input.
- Reference year: It is not always possible to get actual data from the reference year (for MSW data) which was input on the first input screen. Therefore the reference year for the accordant indicator has to be filled in.

National indicator values are displayed for all countries (except Macedonia) which can be selected on the first form. These values can be used as default for the urban indicators (if data are missing) or as recommended value for the national indicators. The database derives from international organisations, such as the United Nations (UNECE, UN-Habitat, UN-ESA), the OECD, the European Union and the World Bank.

Urban indicators

The user's input of these urban indicators is strongly recommended to enable to consider regional peculiarities. In most cases the national default values of these indicators are displayed on the screen. These national defaults are appropriate proxies for the urban values, but should be used only, if urban indicators are not available.

These indicators are usually available from the municipal statistical offices of major cities and are defined as following:

- 'Population aged 15 and 59 years': Percentage of persons between 15 and 59 years of the total city population.
- 'Average household size': Average number of members in a household.
- 'Urban infant mortality rate': The number of infants, out of every 1,000 babies born in a given year, who die before reaching age of one. The lower the rate, the fewer the infant deaths, and generally the greater the level of health care available in a country.
- 'Urban life expectancy': The average number of years newborn babies can be expected to live based on current health conditions. This indicator reflects environmental conditions in a country, the health of its people, the quality of care they receive when they are sick, and their living conditions.

National indicators

These values are usually offered by the tool. If not, data from national or international statistical offices should be used. In this case, the necessary national indicators are to be collected which are defined as following:

- 'Gross domestic product per capita': An aggregate measure of production equal to the sum of the gross values added of all resident institutional units engaged in production. It is here defined in US-Dollar Purchasing Power Parities at 1995 prices.
- 'Infant mortality rate': (The same definition, like for the city indicator.)
- 'Labour force in agriculture': Percentage of persons working in the agriculture as a percentage of the civilian labour force (total labour force excluding armed labour force).

Socio-economic trends

An accurate estimation of the future waste generation has to be based on the development of a city or region. Therefore projections of the underlying indicators are to be used.

Figure 38 shows the data for the indicators which are necessary for the projection of the socio-economic trends. First of all it is very important to use realistic projections of the total number of inhabitants expected in the assessment year. All other input fields refer to indicators which have been defined above in the chapter 'Actual socio-economic conditions'. If available city-related projections should be provided.

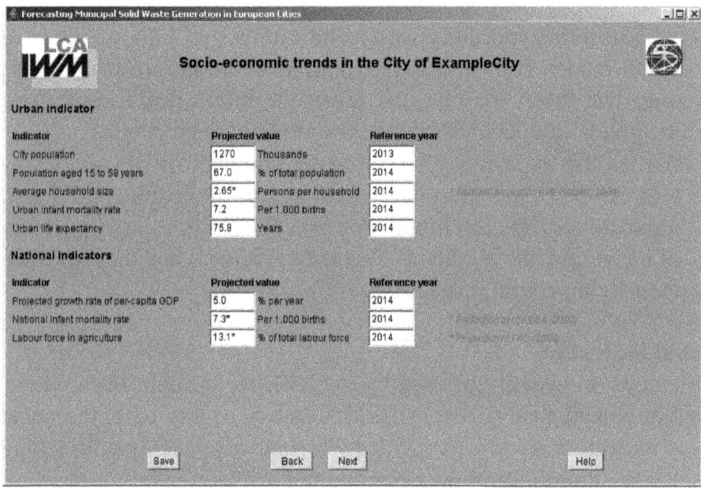

Figure 99. Socio-economic trends.

The displayed default values are based on national projections of international organisations, e.g. United Nations, World Bank, European Union. Depending on the indicator, the horizon of these projections ranges from the year up to 2020. It is recommended to use these defaults only for the national projections. Projections of urban indicators should be based on projections or estimations of the competent regional statistical office.

Similar to the actual indicator values, for each indicator the projected value as well as the future year has to be input.

Waste prevention measures on municipal level

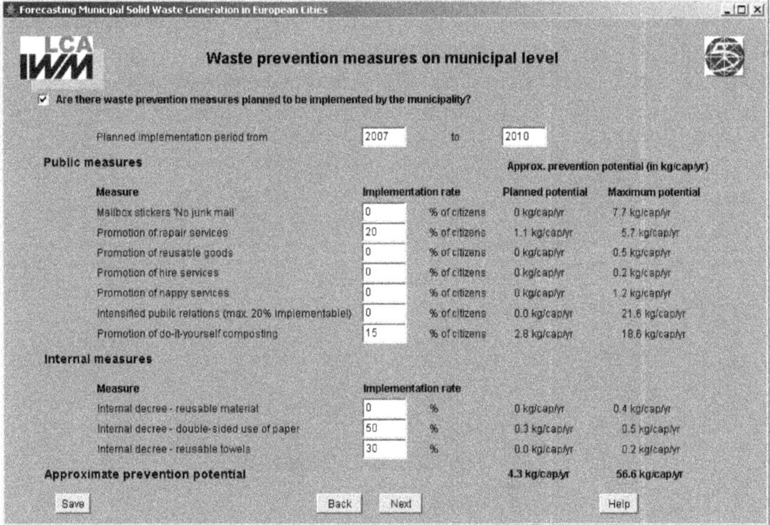

Figure 100. Waste prevention measures on a municipal level.

Waste prevention is defined as quantitative reduction of overall municipal solid waste quantities. Prevented waste quantities do not affect the municipal waste collection and treatment and thus do not cause any environmental burdens in the waste management system. The matter of interest in this project are only measures on municipal level which are planned to be implemented after the reference year.

Ten measures with partly low, but verified quantitative effects are proposed here. It should be determined by the user, in which extent they are planned to be implemented in a defined period of time.

The implementation rate concerns only future prevention effects. If a measure has been implemented fully in the past, then zero has to be input. It is realistic and useful to plan an implementation period with a length of two to six years for the defined set of measures.

As an orientation help, for each measure as well as total, the approximate prevention effect in the reference year, is displayed on the screen.

Saving of inputs

On each input form except the first one the input can be saved, e.g. in the scenario folder. If a new scenario has to be input, the existing inputs can be overwritten and saved in an other file. Alternatively the inputs can be reset by clicking the button 'Reset all inputs' in the first form.

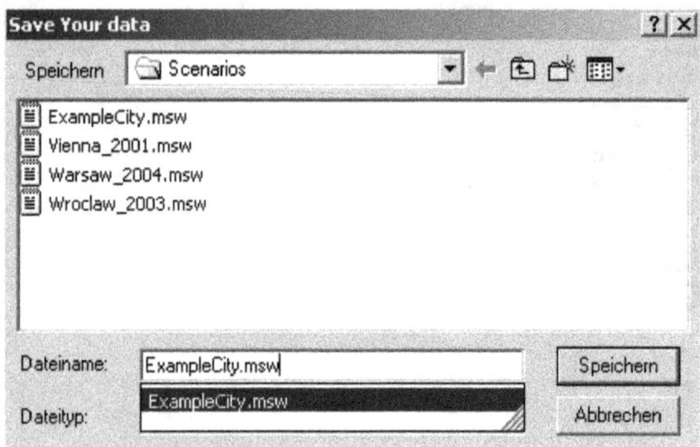

Figure 101. Saving your scenario.

12.3.3 Obtaining the results

The results cover the forecasts with regard to:

- **Municipal solid waste generation**: The quantity of each main fraction is presented (as table or in figures) irregardless if collected separately or as part of the residual waste. These outputs are an essential input for the LCA-IWM Assessment Tool.

- **Separate collection forecast**: Here the collection quantities of all waste types are estimated under different assumptions with regard to the separate collection performance.

All numerical results can be easily exported as "csv" (comma separated value) – file which will be created when clicking on the button "Save as .csv file". Four international formats for different column and decimal separators can be specified for exported data.

Municipal solid waste generation

Figure 102 shows the estimations of the municipal solid waste generation in the assessment year. They are available in tons per year, in kilograms per capita and as mass percentage of total MSW.

The fourth column (tons per year in the assessment year) has to be used as input for the MSWMS Assessment Tool.

Although this tool has been proved to allow a high forecasting accuracy (approx. 4 percent median error for forecasts over 10 years), the accuracy of these forecasts cannot be legally guaranteed within the framework of this project. This tool should be used as a well-funded orientation help.

It is useful to compare the outputs of different realistic scenarios. The existing baseline scenario can for example be related to an optimistic and a pessimistic scenario with a higher/lower economic growth rate, lower/higher infant mortality rate and so on. For this purpose three applications could be simultaneously opened and compared.

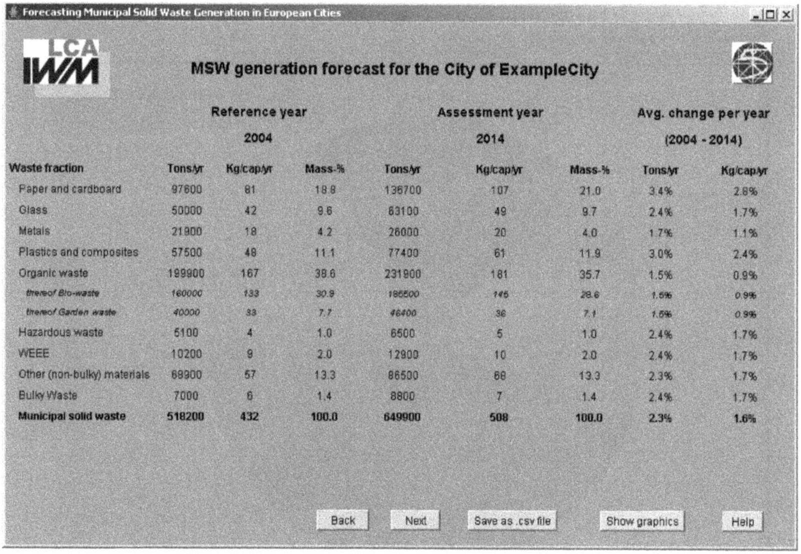

Forecasting Municipal Solid Waste Generation in European Cities

LCA IWM

MSW generation forecast for the City of ExampleCity

	Reference year 2004			Assessment year 2014			Avg. change per year (2004 - 2014)	
Waste fraction	**Tons/yr**	**Kg/cap/yr**	**Mass-%**	**Tons/yr**	**Kg/cap/yr**	**Mass-%**	**Tons/yr**	**Kg/cap/yr**
Paper and cardboard	97600	81	18.8	136700	107	21.0	3.4%	2.8%
Glass	50000	42	9.6	63100	49	9.7	2.4%	1.7%
Metals	21900	18	4.2	28000	20	4.0	1.7%	1.1%
Plastics and composites	57500	48	11.1	77400	61	11.9	3.0%	2.4%
Organic waste	199900	167	38.6	231900	181	35.7	1.5%	0.9%
thereof Bio-waste	*160000*	*133*	*30.9*	*185500*	*145*	*28.6*	*1.5%*	*0.9%*
thereof Garden waste	*40000*	*33*	*7.7*	*46400*	*36*	*7.1*	*1.5%*	*0.9%*
Hazardous waste	5100	4	1.0	6500	5	1.0	2.4%	1.7%
WEEE	10200	9	2.0	12900	10	2.0	2.4%	1.7%
Other (non-bulky) materials	68900	57	13.3	86500	68	13.3	2.3%	1.7%
Bulky Waste	7000	6	1.4	8800	7	1.4	2.4%	1.7%
Municipal solid waste	**518200**	**432**	**100.0**	**649900**	**508**	**100.0**	**2.3%**	**1.6%**

Back | Next | Save as .csv file | Show graphics | Help

Figure 102. Results table to the MSW generation forecast.

Figure 103 and Figure 104 show other examples of graphical presentations of the results.

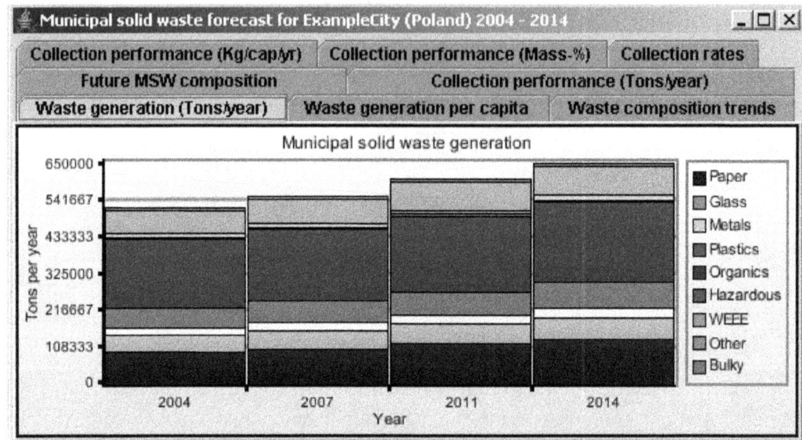

Figure 103. Municipal solid waste generation forecast in tons per year.

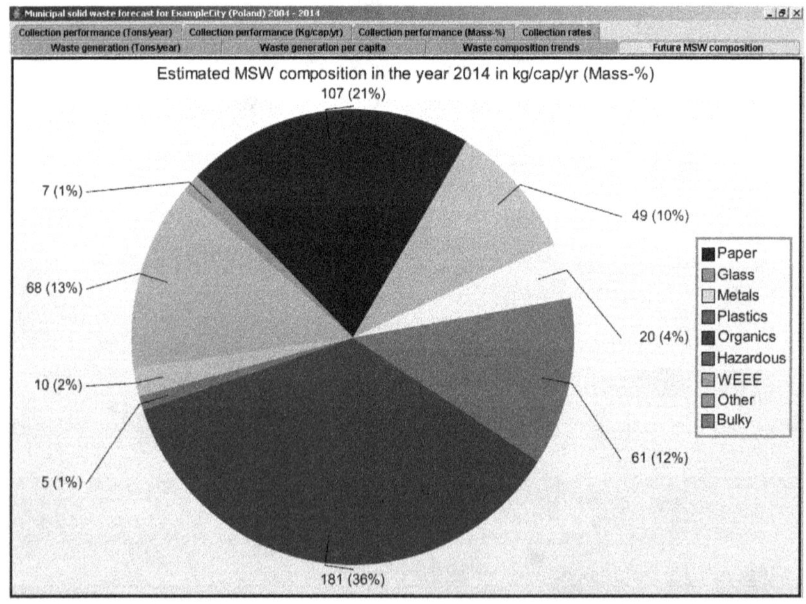

Figure 104. Future MSW composition as per capita quantity and mass percentage.

Separate collection scenarios

The future collection quantities are a matter of interest in waste management. The separate collection quantities strongly depend on the ability and willingness of the municipality to collect a number of fractions separately.

Figure 105 shows the quantities in the case of two future scenarios:

- **Steady collection rates ('Status quo')**: It is assumed that the same percentage of each waste fraction will be collected in the assessment year (e.g. xx% of the total waste paper).
- **Achieving 'target rates'**: Here it is assumed to achieve target rates for the collection of paper and cardboard, glass, plastics and composites and organic waste.

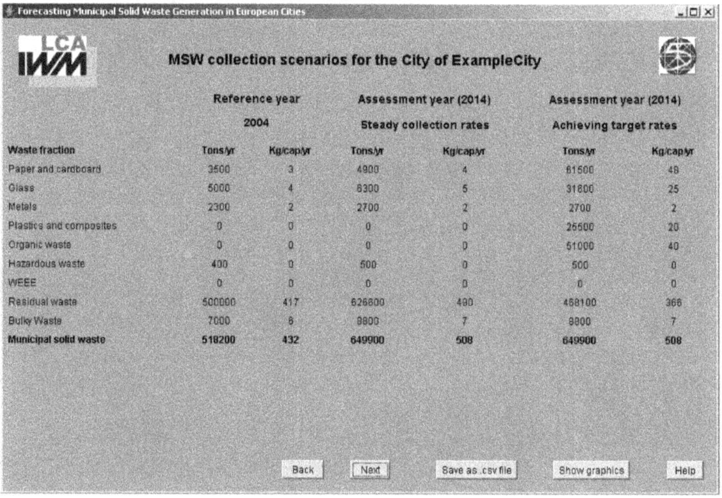

Forecasting Municipal Solid Waste Generation in European Cities

LCA IWM

MSW collection scenarios for the City of ExampleCity

	Reference year 2004		Assessment year (2014) Steady collection rates		Assessment year (2014) Achieving target rates	
Waste fraction	Tons/yr	Kg/cap/yr	Tons/yr	Kg/cap/yr	Tons/yr	Kg/cap/yr
Paper and cardboard	3500	3	4900	4	61500	48
Glass	5000	4	6300	5	31800	25
Metals	2300	2	2700	2	2700	2
Plastics and composites	0	0	0	0	25500	20
Organic waste	0	0	0	0	51000	40
Hazardous waste	400	0	500	0	500	0
WEEE	0	0	0	0	0	0
Residual waste	500000	417	626800	490	468100	366
Bulky Waste	7000	6	8800	7	8800	7
Municipal solid waste	**518200**	**432**	**649900**	**508**	**649900**	**508**

Back | Next | Save as .csv file | Show graphics | Help

Figure 105. MSW collection scenarios.

Figure 106 and Figure 107 show the graphical results to the separate collection performance. Additionally the scenario 'Optimum value' (Assuming very good collection performance) is shown here.

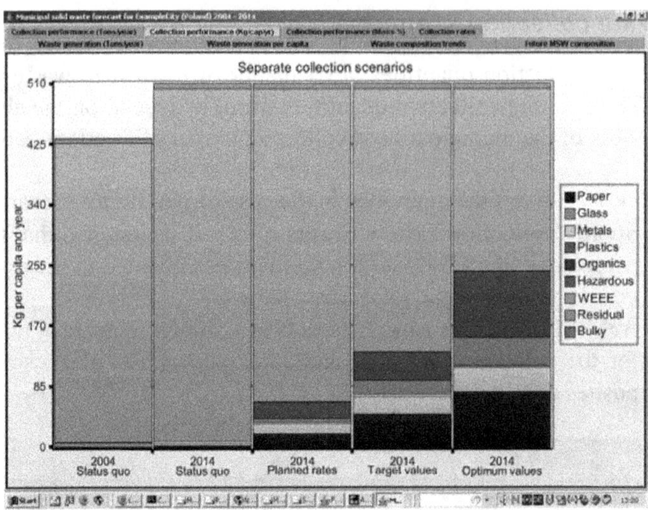

Figure 106. Separate collection scenarios as per capita quantities.

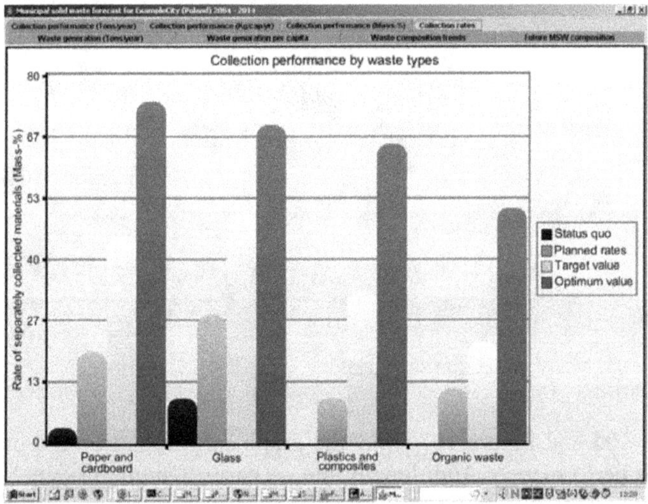

Figure 107. Separate collection performance – collection rates.

Additionally to these collection scenarios, the planned separate collection rates can be input for every separately collected waste stream. The resulting future waste streams are also available in tabular and graphical format.

13 MSWMS Assessment Tool guide

13.1 Introduction

This documentation is designed to explain the main features and basic functionality of the LCA-IWM Assessment Tool. Examples are given with the purpose of illustrating the main functionalities.

In preparing this guide, it has been assumed that the user has a basic familiarity with spreadsheet calculations and is familiar with LCA-IWM documentation, which describes the theoretical basis for the model. The material included in the LCA-IWM reports is not repeated in this manual.

This study has tried to design a visual basic interface which allows the user to load data, operate and view the results' calculation with a minimum understanding of Excel, and without any knowledge of Visual Basic. Nonetheless, the more advanced Excel user will be able to master the operations of the LCA-IWM Assessment Tool more rapidly and be more comfortable making their own modifications to the program execution.

13.1.1 What is the LCA-IWM Assessment Tool?

The tool is an Excel interface which allows the user to assess, compare and improve the environmental, economic, social sustainability of the waste management system of a city or region either existing or hypothetical.

LCA-IWM Assessment Tool applications are:

- To support the planning and monitoring of the waste management system of a municipality.
- To enable local politicians introducing new and improving existing waste management systems.

13.2 Getting Started

13.2.1 Installing LCA-IWM

Download the Assessment Tool from the LCA-IWM project website www.lca-iwm.net.

Once the user has downloaded 'LCA-IWM Assessment Tool' (compressed archive), he gets a zip file. When the user unzips the tool file then gets a LCA-IWM folder in his computer. In this folder the user finds four folders called: modules, scenarios, assessment session, output and one file called main (Figure 94).

The type, contents and purpose of each file are as follows:

- *'modules folder'*; all the Excel files used in the tool calculation are located in this folder. These worksheets are normally hidden from view. The user does not to see the calculations. However if he would like to check some calculation he can have a look at the Excel modules.

- *'scenarios folder';* location of the case studies which the user creates. The tool locates all the scenarios in this folder by default. It is important not to locate the scenarios in another place.

- *'assessment session folder'*; location of the assessment studies which the user defines. An assessment session is the tool way to see results. A session can be formed from 1 to 4 scenarios. The tool always locates the assessment session in this folder. This location must not be modified.

- *'output folder'*; location of the output files. The assessment session creates an output file, which is always called the same as the assessment session. In this way the user always receives an output file which can be used for further reporting purposes

- *'main.xls'*; is the VISUAL BASIC interface. By double clicking in this file the user opens the tool.

Figure 108. LCA-IWM tool folder

ATTENTION: the user is not allowed to modify the tool folders names

To open the tool, select the file 'main' from the LCA-IWM tool folder by double clicking. When the user opens the file 'main', the user opens the tool. The tool screen is organised as three sheets:

- *'MAIN sheet';* is the Assessment Tool presentation
- *'input sheet';* allows the user to define the scenario, by defining the variable input values
- *'output sheet'*; allows the user to view the assessment session, meaning the results.

Also the user gets three new items on the 'Excel menu bar' (SILCA-IWM, treatments plants and output) (Figure 109). These three items allow the user to define the system by means of entering the input. The buttons in the input screen may be used for the same purpose.

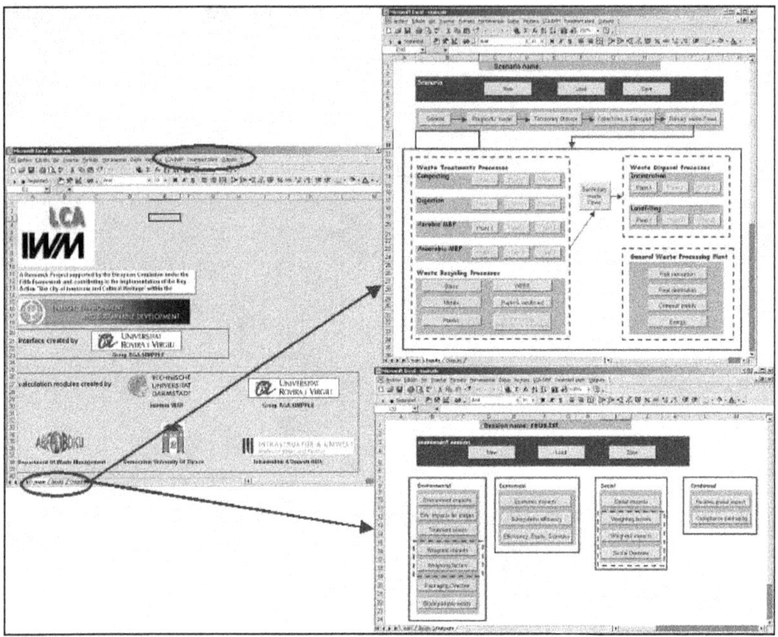

Figure 109. LCA-IWM Assessment Tool initial screen.

13.2.2 Tool skills

13.2.2.1 Macros activation

Before opening the tool, the user should activate the macros in his Excel software; go to the 'tool menu', select 'Macros menu', and then go to the menu entitled 'Security menu' and select 'Choose the security level'.

Excel --> tool --> macros --> security --> Choose the security level

Please note the following options:

- *'medium';* every time that the user opens the main file, the user gets a question; do you want to activate your macros or not? The user should answer that he wants to activate the macros.
- *'low';* macros are activated automatically.

13.2.2.2 Regional number configuration

The LCA-IWM Assessment Tool was designed with the following fixed 'Regional configuration' number:

- Decimal symbol "." (e. g. 0.15)
- Thousand symbol of separation "," (e. g. 1,000,000).

If the regional configuration number is different than specified above, it may lead to mistakes in the calculations of the results.

Pay attention to your computers regional configuration number.

13.2.3 Computer Requirements

There are minimum hardware and software requirements which must be satisfied to ensure that the product functions properly.

HARDWARE: Pentium III 500 Mhz
Hard-drive storage 50 MB

SOFTWARE: Windows 2000 Office 2000
Windows XP Office XP

Programming language:
Visual Basic for applications v.6.

13.3 Using the LCA-IWM Assessment Tool

13.3.1 Opening the LCA-IWM Assessment Tool

When the user opens the tool, a dialog box appears on the screen and prompts him to either (Figure 110):

- create a new scenario
- create a new assessment session
- load a scenario
- load an assessment session

The user chooses whether he wants to create a new LCA-IWM working file (scenario or assessment session) or to load an existing file.

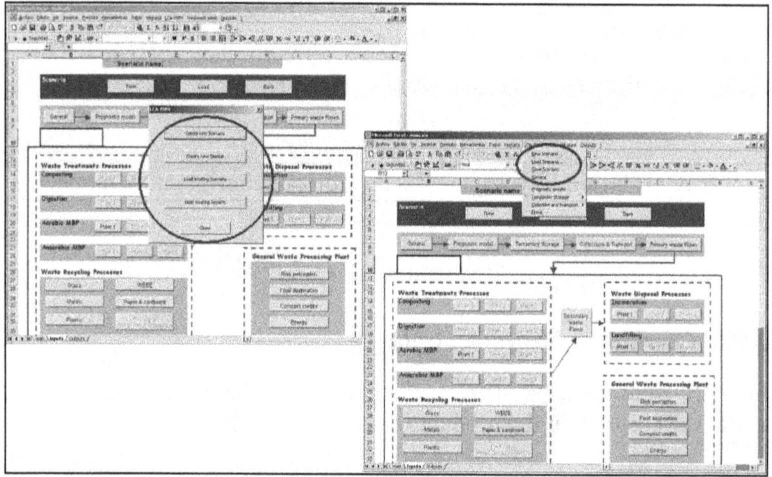

Figure 110. New scenario creation.

13.3.2 A scenario

A 'scenario' is a case study which the user defines entering the inputs variables values.

The tool allows the user three actions with a scenario:

- 'Create a scenario'

- 'Load a scenario'
- 'Save a scenario'

There are three different ways to create and load a scenario:

- When the user opens the tool the user gets into the dialog start up menu showing two options; 'create a scenario' and 'load a scenario'.

- In the 'LCA-IWM Excel' menu bar there is a button called 'new scenario', 'load scenario' and 'save scenario'.

- In the input screen, there is a button called 'new scenario', load scenario' and 'save scenario'.

ATTENTION: Every time the user modifies something in the scenario, it has to be saved. If the scenario is not saved, the modifications are lost.

The user may create as many scenarios as he wishes.

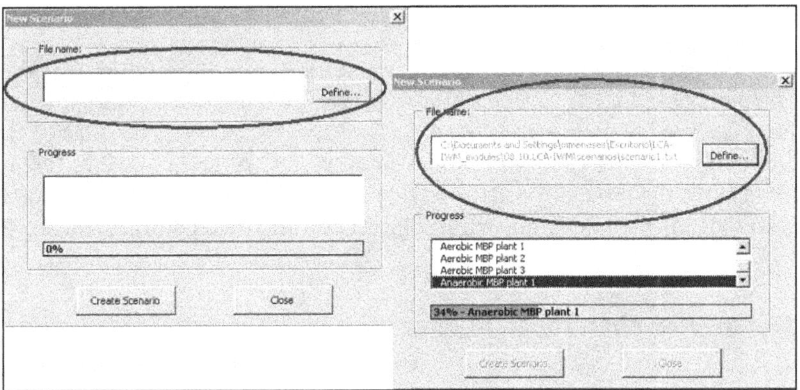

Figure 111. New scenario screen.

Once the user has created a new scenario, he is asked to name and save (= locate) the file in the default directory ('scenarios folder'). The dialog box (Figure 111) appears on the screen to enter this information.

The scenario file can be given any eight-character name followed by '.xls'. Characters which are not allowed as parameter names in Excel, such as, blanks, commas hyphens, plus signs, etc., and must NOT be used in this name (consult the Excel manual for a full list of restricted characters). The underline character '_' and numbers are allowed in this name.

By default a 'new scenario' is located in the tool folder called 'scenario'. This folder is created by default when the user installs the tool in his computer. This is important because the scenario should always be located in the 'scenario folder' in order to avoid problems.

After naming and locating the case study the user creates the new scenario. In this moment all the Excel menus which the tool needs to assess the system are loaded. In the 'new scenario menu' the user can see the file which is loaded, its progress and the percent of files loaded.

If the user loads or saves a scenario, the box dialog menu is similar to the 'new scenario dialog', the user defines the scenario which he wants to load, and he can see the progress (the percentage of files loaded). When the user 'saves a scenario' the action is exactly the same. Saving and loading a scenario takes only a few minutes to be completed. The download progress windows informs the user of the status of the installation.

If the user selects the option of loading or saving an existing file, the user will be asked to select a file from the LCA-IWM directory. A file from the outside "scenario LCA-IWM directory" should not be selected or else the macros will not work, since moving outside the LCA-IWM directory causes the directory path to be reset. If the file the user selects is not a worksheet file or does not match the LCA-IWM worksheet template, it will be rejected.

13.3.3 Flow chart tool

Once a new scenario has been created or an old scenario has been loaded the 'LCA-IWM Assessment Tool' is ready to carry out calculations.

The tool input screen shows the typical flow of tasks in a LCA-IWM calculation and provides buttons which operate macros. The user can define the system, by means of entering the input variables. The flow chart is illustrated by means of arrows in the inputs screen which will help the user through the process (see Figure 112). Depending on the waste fractions and also on the treatment plants defined the treatment plants and the waste recycling process appear activated or deactivated automatically.

Each updating of an existing scenario can be used to include and exclude a waste management stage (temporary storage, collection & transport, treatment plants) or a sustainability aspect (environmental, economic, and social), also the user can always update the treatment plant considered in the system. The user has complete flexibility to select or not to select a pathway to be included in a calculation.

There are essential variables which should be defined: number of city inhabitants and waste amount. If these variables are not defined, the system has no logical sense and because of that the tool will not work.

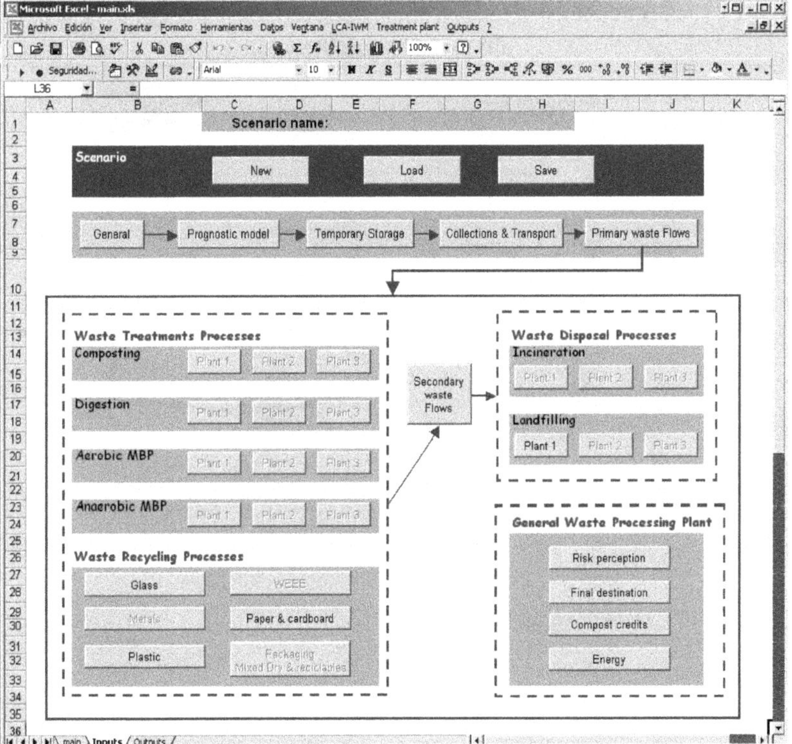

Figure 112. Flow chart diagram.

ATTENTION: It is obligatory to define city inhabitants & waste amount.

13.3.4 Default value

Most of the variables have a default value which the user can see by passing the cursor over the cell. The default value appears as a message (see Figure 113).

The user gets this message, and then has to type the value in the cell. This is so that the user becomes conscious of the value which is being introduced. The user can enter this value or another which he considers more appropriate.

ATTENTION: If no variable value is entered, the calculation is done with the default variable value.

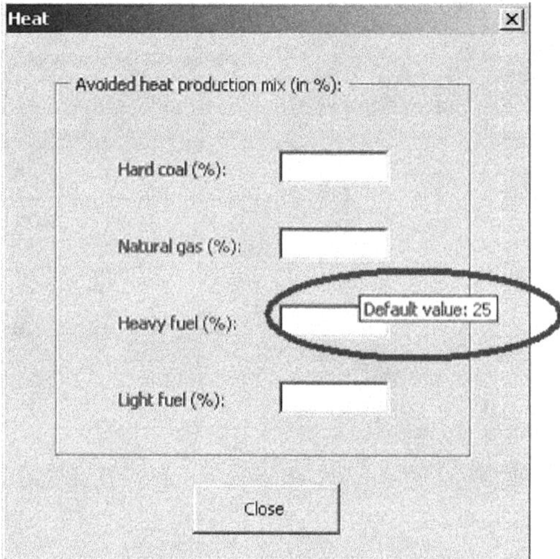

Figure 113. Default variable values.

13.3.5 Help variable messages

Most of the variables have 'help messages' which the user can see by clicking F1 when the cursor is on a cell (see Figure 114). The objective of these 'help messages' is to clarify confusing input variables as well as tool functionality.

The help messages start with three possible symbols;

- NN; input is not necessary (e. g. can be given on a higher/other level of detail as well)
- D; default value is available and will be taken if no other values are entered
- O; input is obligatory

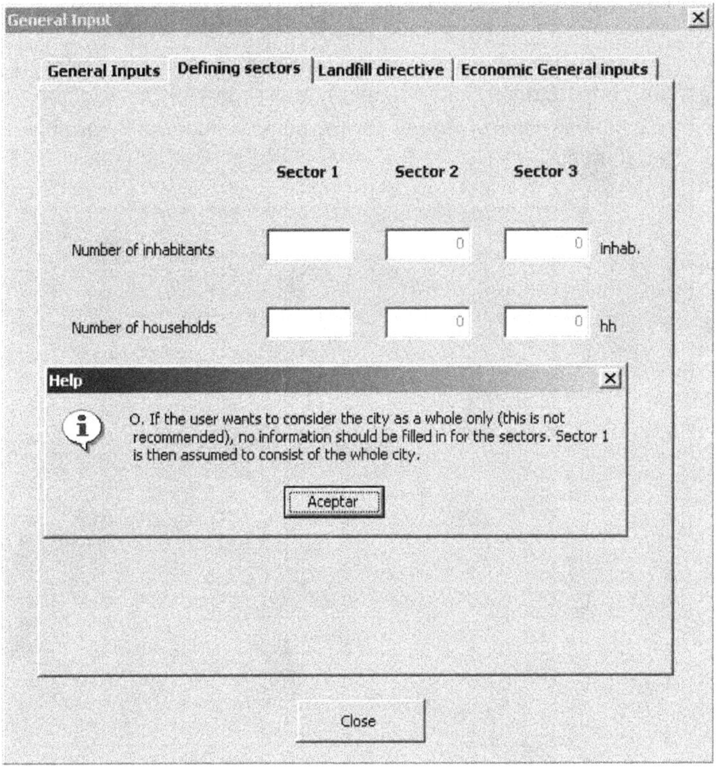

Figure 114. Help variables messages.

13.3.6 LCA-IWM Main Menus, Contextual Menus and toolbars

13.3.6.1 General input

The first step is to define the general input dialog menu (see Figure 115) this dialog menu box is divided into four sections or windows:

- '*General inputs*'; in this dialog menu the user defines general city characteristics.
- '*Defining sectors*'; the tool allows the division of the city into three sectors. The user has to define the general sector properties (e.g. inhabitants, area …). If the user wants to consider the city as a whole only no information should be filled in for the sectors.
- '*Landfill directive*'; the waste fractions amount generated in 1995 are defined, regardless whether it was separately collected or not.
- '*Economic general inputs*'; some general economic variables inputs are defined.

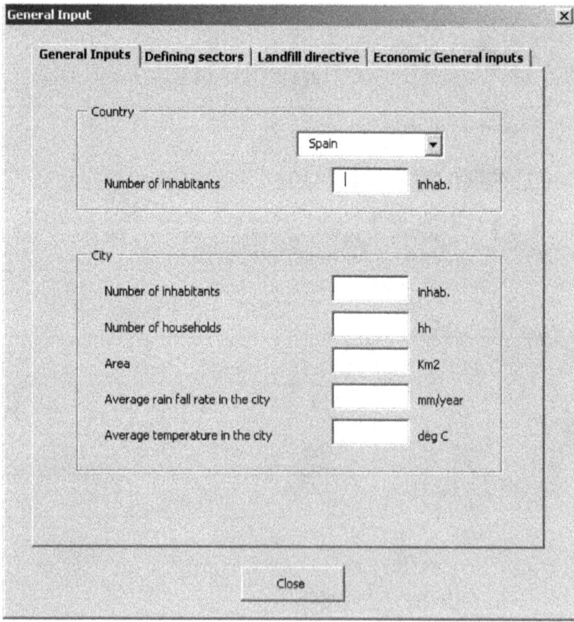

Figure 115. General input dialog menu.

13.3.6.2 Prognostic results menu

In this menu the user enters the results from the 'prognostic model'. It is sort out into two menus; 'current year' and 'prognostic year', respectively. The waste fraction generated for each fraction in both the current and the prognostic year are entered in this menu.

The idea is that there is a year X and a year X + 10, which are called current year and year of assessment. This last comes from the prognostic model, which needs the current waste data to calculate and forecasts the waste data in the year of assessment.

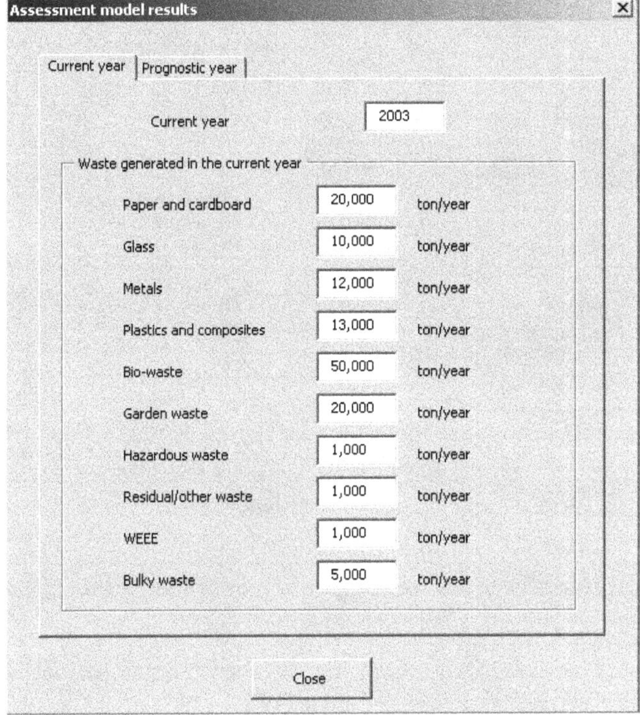

Figure 116. Prognostic model menu result

13.3.6.3 Temporary Storage

This section defines the Temporary Storage of the waste management system. The user has to define the environmental inputs, the economic inputs and the social input variables. The whole tool has the same structure for all waste management stages. In each stage the user has a dialog box for the environmental input, another for the economic inputs and another for the social input. On the left corner in the dialog box, the tool always indicates the waste management stage and the kind of input variable (see Figure 117).

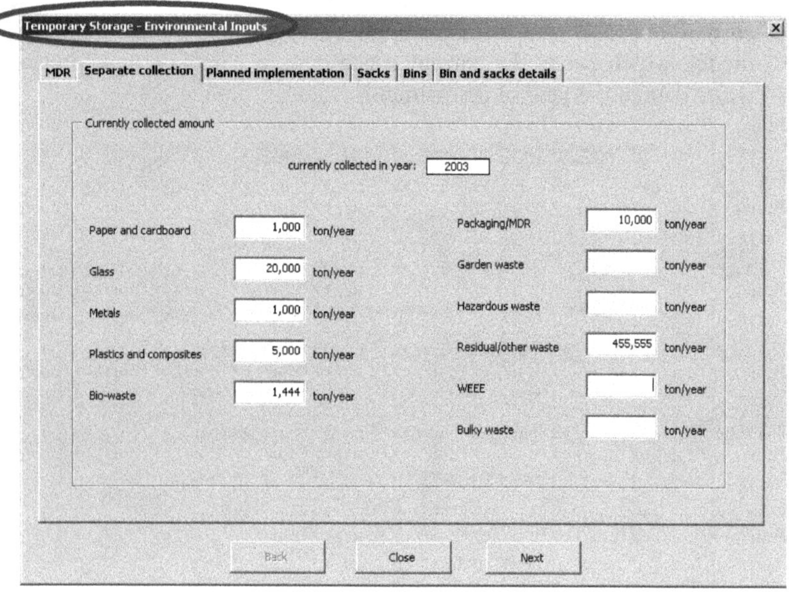

Figure 117. Temporary storage – environmental inputs.

Environmental inputs of Temporary Storage menu

This menu is sorted out into 6 dialogs:

- *'separate collection';* the user enters the amount of current waste collected (see Figure 117).
- *'planned implementation'*; To *know* how much waste will be collected separately depends on how long the fraction is already collected separately. Therefore it is possible to give in these different years of implementation depending on the waste fraction. The collection targets to be achieved can be either 'target values' (average of European cities belonging to the highest pros-

perity group) or ‘optimum values’ (highest 10% percentile of European cities belonging to the highest prosperity group). The users has to choose which values are assumed to be reachable in a period of 10 years after implementation of the separate collection. If the current system already reaches values which exceed these targets, these current results are taken as future targets.

If in any scenarios a fraction is separately collected in the current year the user chooses: currently existing. If it is planned for later (logically somewhere before the year of assessment) the user chooses that year.

In the 'current' scenario one would rather not plan any other fractions than there are at the moment in the city, since one wants to make a baseline scenario. For the other, one can choose to implement the separate collection at any time (see Figure 118).

A sector cannot introduce a selective collection of some fraction in part of the sector, e.g. selective collection of biowaste for 50% inhabitants of new multi-houses sector. The sector can be defined with the multi-houses where one fraction is either collected separately or not.

- *‘sacks’, ‘bins’, ‘bins and sacks details’;* in these three dialogs, the user defines the features of the sacks and bins. In a city the sacks and bins could have different systems depending on the area, because of that it could be defined depending on the city sector. Per city sector and fraction the user can select one volume of sacks, one volume/material of small bins and one volume/material of large bins.

- *‘MDR’;* in this dialog the user defines if the waste management system has ‘mixed dry recyclable faction’ and also the waste factions which form it.

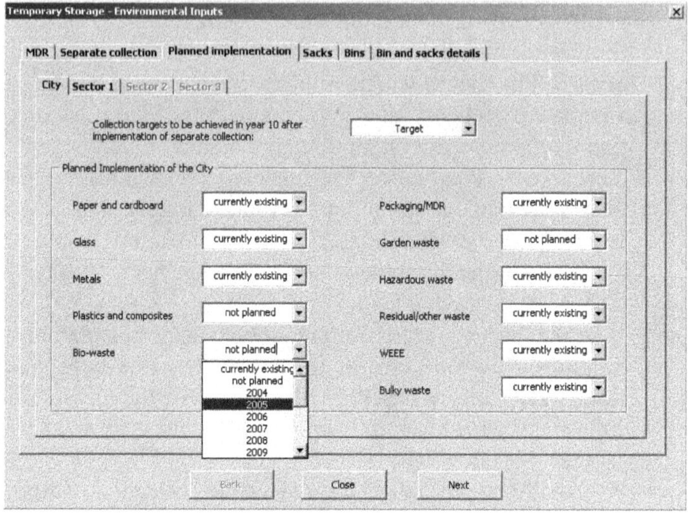

Figure 118. Planned implementation of separate collection.

Economic inputs of Temporary Storage

The economic inputs menu of the temporary storage is divided into three sections (Figure 119):

- *'location cost of bins';* the cost over and above the purchase cost for distributing the bins around the city.

- *'end of life cost';* as the percentage of the purchase costs. Negative value indicates a positive rest value of bins (at the end of their life time).

- *'maintenance cost';* as the percentage of the equivalent annual discounted total purchase cost of the bins.

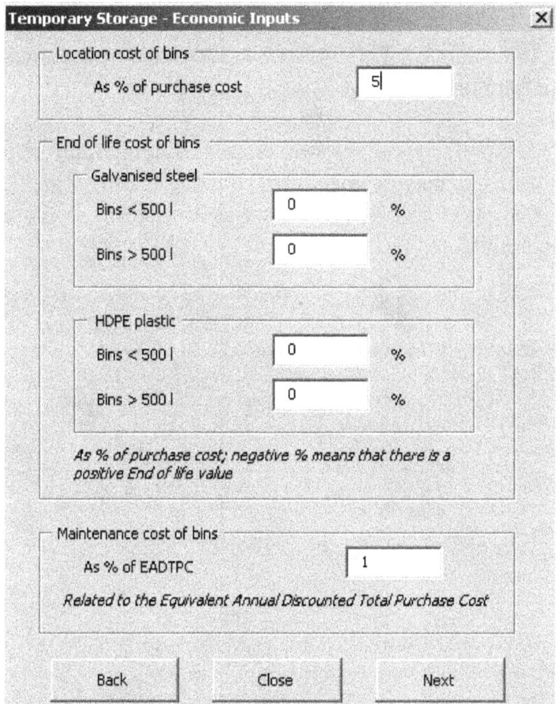

Figure 119. Temporary storage - economic inputs

Social inputs of Temporary Storage

In the Temporary Storage the social impacts evaluated are (see Figure 120): odour, visual impact, convenience, private space, noise, complexity and distribution and location.

- *'Odour';* The odour fractions considered by the tool are; paper, glass, metals, plastics and composites, bio-waste, packaging/MDR and residual /mixed + garden waste, bulky waste, hazardous and WEEE. The user input is only necessary for the last 4 waste fractions because for the other ones it is already available from another model part.

- *'Convenience'*; This indicator calculates how convenient the total temporary storage system is for the public.

- *'Noise';* Disposal of waste into a temporary storage causes noise. The user defines relative noise factors which enable a differentiation between various waste fractions, container materials and container sizes.

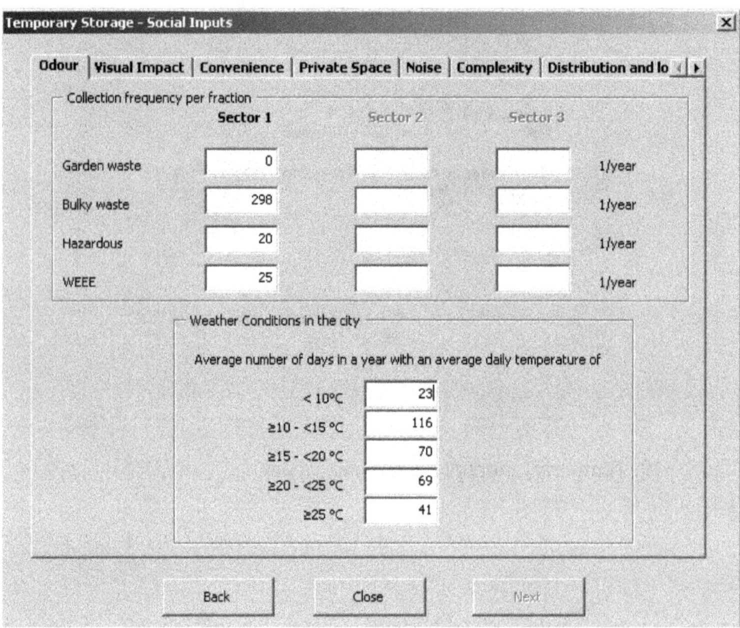

Figure 120. Temporary storage – social impacts

13.3.6.4 Collection and Transport

Environmental inputs of Collection and Transport

The user has to describe the environmental, economic and social properties of the 'Collection and Transport stage'. This menu is organised into six main parts (see Figure 121):

- '*General*'; general collection and transport distance.

- '*Transfer station*'; the user identifies for which waste fraction a transfer station is used, and also in which sectors.

- '*Distance*'; this section describes the distance in kilometres from transfer station to facility and from sector to facility.

- '*Vehicles used*'; the different kinds of vehicles used are identified as well as their characteristics (see Figure 121). The user enters the capacity by fraction of the trucks. If a vehicle has a design capacity of 12t for certain materials the maximum capacity is smaller. E.g. if the user transports polystyrene in a 40 tons truck, the user will only be able to load about 1t. Light packaging (metals, plastics, MDR, default values is 25%) is not assumed to use the maximum capacity of the truck, since the materials are lighter (in contrast to paper, residual and bio-waste default values is 100%).

- '*Vehicles type*'; are divided to vehicles for transport and for collection. Both of these are available for bins < 500 litres and sacks, and bins > 500 litres.

- '*Performance data'*; this section describes aspects about the collection and transport, as average distance and average speed between two containers, number of collector per crew, etc.

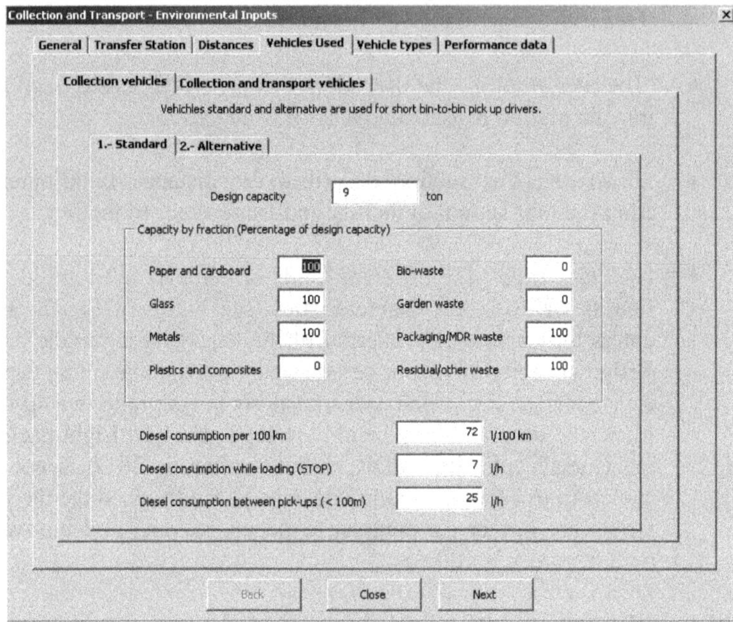

Figure 121. Collection & transport – environmental input.

Economic inputs of Collection & Transport

This section defines the economic aspects for the Collection and Transport stage for the waste management system. It is divided into three sections: personal cost, vehicles cost and overheads (Figure 122).

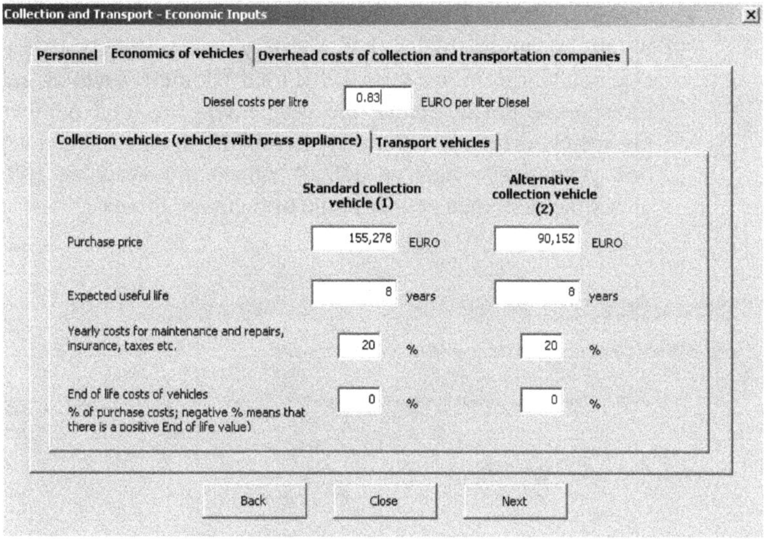

Collection and Transport - Economic Inputs

Personnel | Economics of vehicles | Overhead costs of collection and transportation companies

Diesel costs per litre 0.83 EURO per liter Diesel

Collection vehicles (vehicles with press appliance) | Transport vehicles

	Standard collection vehicle (1)		Alternative collection vehicle (2)	
Purchase price	155,278	EURO	90,152	EURO
Expected useful life	8	years	8	years
Yearly costs for maintenance and repairs, insurance, taxes etc.	20	%	20	%
End of life costs of vehicles % of purchase costs; negative % means that there is a positive End of life value)	0	%	0	%

Back Close Next

Figure 122. Collection & transport – economic input.

Social inputs of Collection & Transport

The social impact of the Collection and Transport stage is evaluated by means of 5 indicators; noise, traffic, urban space, employment quality and employment quantity.

- *'Noise'*; in the first field 'general' the user has to enter the 'total length of the road network within the city. Also the relative division into three road kinds for the road network and the road driven on in both collection and transport should be inserted. The required value which is indicating the traffic density is the ChNV, Characteristic number of vehicles per hour for each road type for day and night (the 'background traffic'). Alternatively, the user is able to insert the ADT, Average Daily Traffic deriving from traffic censuses (automatic or sample counting). The ADT value applies for 24 hours, thus it is not divided in day and night cycles. Both values are specific values which may be available at the city's traffic department.

- *'Traffic'*; the user has to enter the total number of km's driven by all transport vehicles (i. e. not only waste related) within the city boundaries in one year. If the total kilometer level of freight transport is not available, the user enters the total number of kilometers driven by all transport vehicles within the country in one year. Some relative default values are available (please check the reference year as some of them are too old).

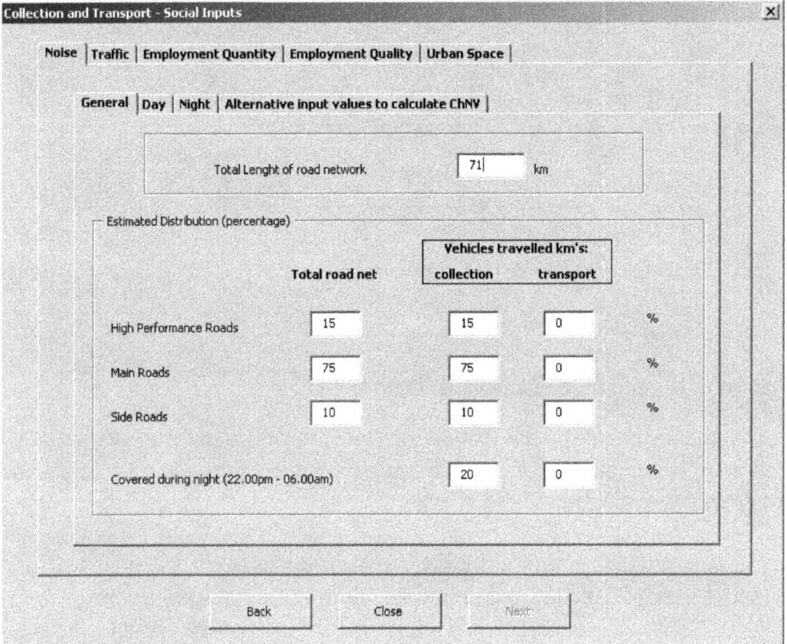

Figure 123. Collection & transport – social inputs

13.3.6.5 Primary flows menu

Flow menu dialog expands the information which the user entered to identify the residual and organic waste flows, allowing the user to define the waste treatment plants as well as the number of treatment plants and the waste amount treated for every treatment plant. This menu is divided into two sections: organic waste and residual waste.

Organic waste

In the Figure 124 the user sees the organic waste flows menu. The tool allows three treatment options to treat organic waste; composting, digestion or both. The user can choose the treatment plant to consider in his system. Also the user defines the number of plants in his city, always keeping in mind as maximum three at a time.

On the upper right corner the user can see the amount of garden and biowaste collected in the system. After selecting the treatment plant, the user defines the waste amounts treated in each plant. If the user does not select a plant it appears deactivated in the main input screen.

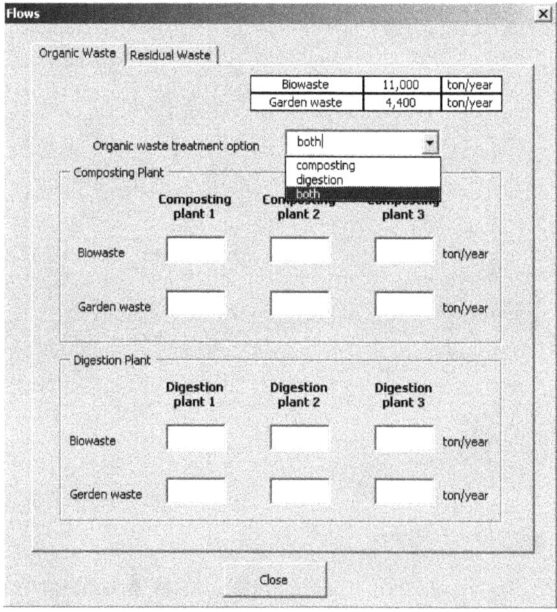

Figure 124. Organic waste flow menu.

ATTENTION: The sum of the garden or biowaste treated cannot be in excess of the amount of these wastes collected in the system. To ensure a proper functioning, waste amounts are always entered, also if relevant default values are provided. The same is valid for all primary and secondary waste input windows.

If the sum of organic waste treated in the composting and digestion plant exceeds the organic waste generated in the system, then the user receives a message, and the organic waste cell changes the colour to red, as the user can see in the Figure 125 and the tool will not be able to calculate.

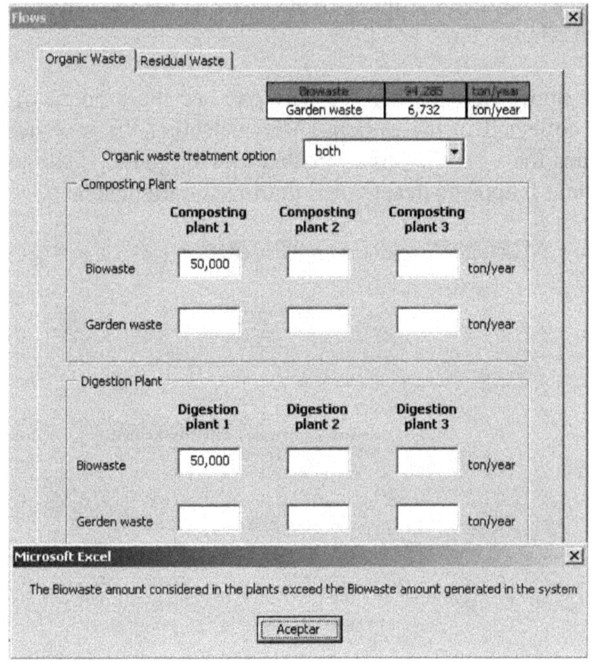

Figure 125. Error message

Residual waste

In this dialog menu box the residual waste treatment is defined (see Figure 126). This screen is similar to 'organic waste flows menu'.

There are four residual waste treatment plants considered here: Anaerobic MBP, Aerobic MBP, Incineration and Landfill. The user chooses the treatment plant used in the system, the number of plants, and the residual waste treated in each plant.

If the residual waste treated in the plants considered exceeds the residual waste generated in the system, then the user receives a message, and the residual waste cell changes the colour to red and the tool will not be able to calculate.

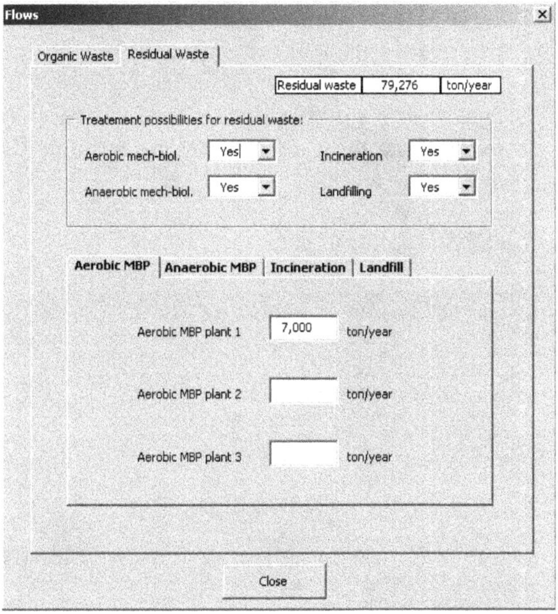

Figure 126. Residual waste flow menu.

In the secondary waste inputs flow menu the product flows of the MBP plants and the rejects of all treatment/recylcing plants are considered (see section 13.3.6.8). If no waste is directly incinerated, but MBP outputs or rejects will be incinerated, the incinerator should be activated.

13.3.6.6 Waste Treatment plants

The tool takes into account four Treatment plants, which are classified depending on the treated organic or residual waste. The system allows the user to have from one to three plants. The user defines the kind of Treatment plant, the number of plants and the amount of waste treated in the flows primary menu.

The waste management treatment plants menu is divided into Figure 127;

- *'Environmental inputs'*;. This menu is classified into three sections;

    - waste amount treated in the treatment plant is indicated. The value is fixed in this menu, it is the result of internal calculations from the tool.

    - inputs for plant operation, if the plant receives waste from other areas, the user specifies the quantity delivered from these other areas.

    - emission controls used are identified.

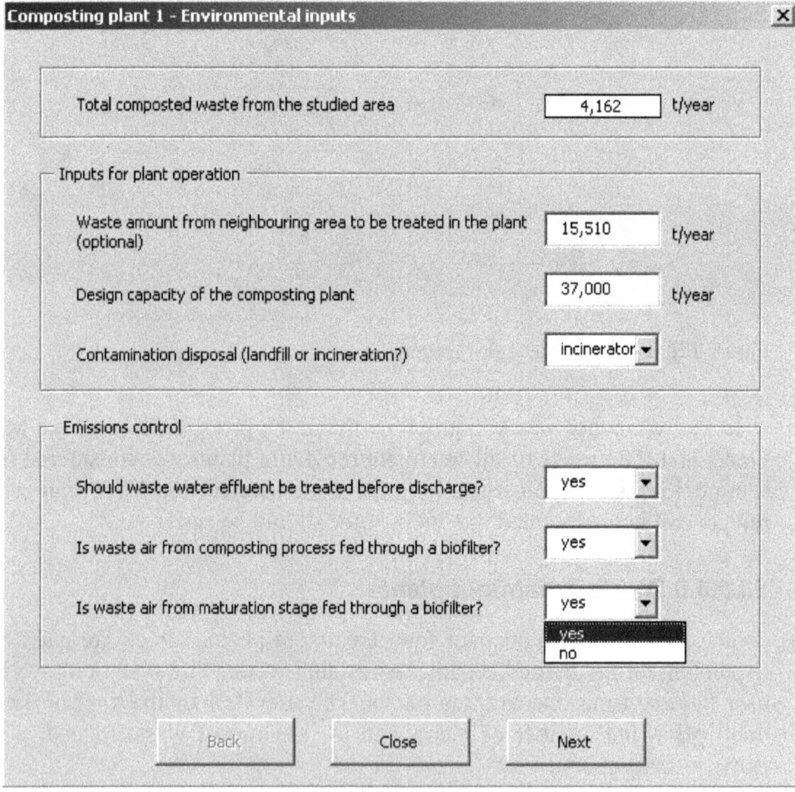

Figure 127. Waste treatment plant- environmental inputs

If in MBP is chosen that the high caloric fraction (HCFr) is not used for energy recovery in a cement kiln, it goes as a 'secondary waste flow' into the Incinerator. So if one does not plan an Incinerator, it is necessary to choose the option of utilisation of HCFr in the cement kiln. If cement kiln was chosen, HCFr flow leaves the MSWMS, but the environmental impacts of its utilisation in a cement kiln are still accounted for. If an MBP is chosen whose the low caloric fraction (LCFr) is not used for land recultivation, it goes as a 'secondary waste flow' into the landfill. If destination of LCFr is 'landfill recultivation', this flow leaves the MSWMS, but the environmental impacts of this application of LCFr are still considered in the tool.

- *'Economic inputs';* this menu is divided into three dialogs (see Figure 128):
    - General
    - Transport vehicles; The user specifies the investment and operational cost for the total transport fleet needed(for the whole plant, including treatment of waste of neighbouring areas).
    - Revenues from secondary materials and energy

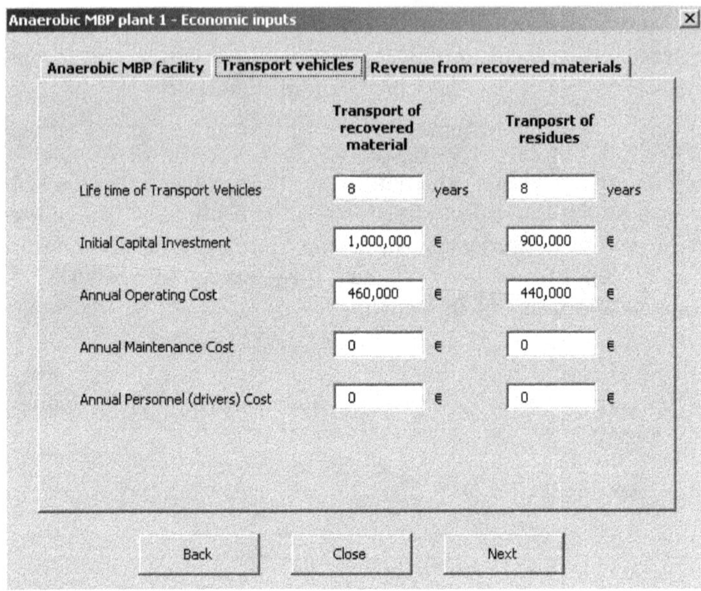

Figure 128. Waste treatment plant – economic inputs

- *'Social inputs'*; is sorted out in five dialogs; odour, visual impact, traffic, employment quality and employment creation Figure 129.

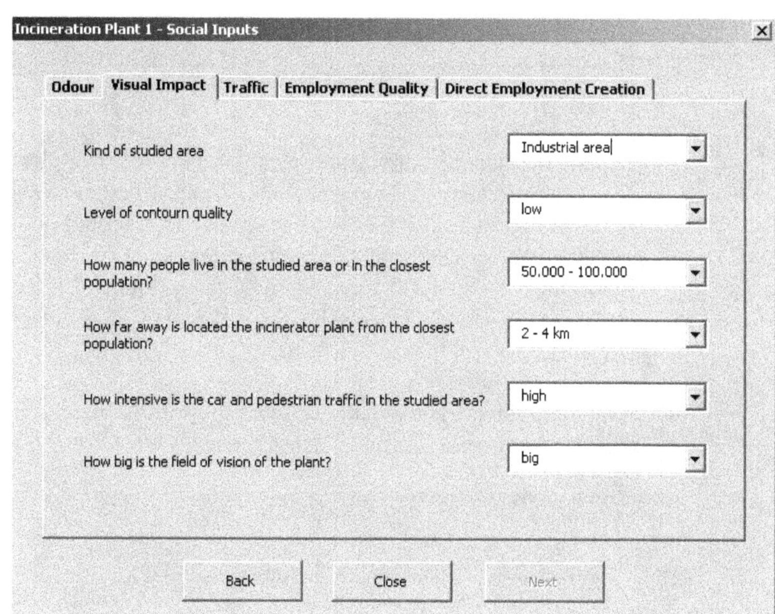

Figure 129. Waste treatment plant – social inputs.

13.3.6.7 Waste Recycling processes

Recovered materials from household wastes which are reprocessed can be used to replace virgin materials, and this may result in overall saving in raw materials and energy consumption and emissions to air, waste and soil. These 'waste Recycling processes' allow balancing, for the different household waste fractions, the environmental advantages and disadvantages of materials recycling processes against virgin materials production processes. The tool assesses six waste recycling processes; glass, metals, plastics, WEEE, paper and cardboard, and mixed dry recyclables.

All separately collected waste flows which enter the Recycling processes are already defined in the Temporary Storage Menu.

All the Recycling processes are divided into:

- *'environmental inputs'* (Figure 130);
  - general; the user identifies the waste fraction collected separately and the composition.

- consumption for sorting; the cleaning and crushing process is defined.

- rejects; the quantity of waste reject and its destination, landfill or incineration are defined. The 'Rejects Factor' determines how much of the initial contaminants is sorted out as rejects. The idea is, that if e.g. the contamination level of the input (of e.g. 1000 tons) is 10%, the amount of rejects could be 150 tons (Rejects Factor = 1.5) but also 50 tons (Rejects Factor = 0.5).

- transport distances; the distance from the sorting plant to the rejected treatment facility is characterised.

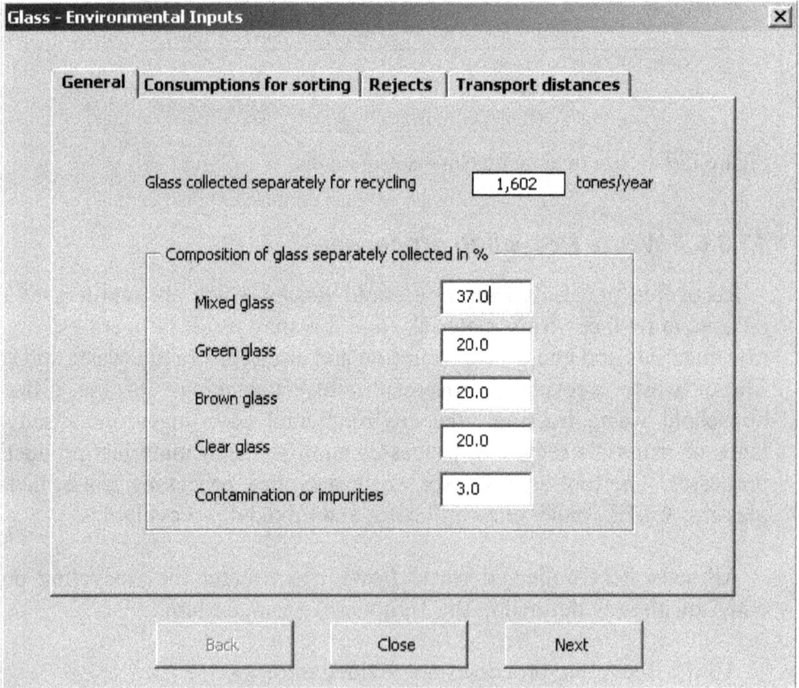

Figure 130. Waste recycling process – environmental inputs

- *'economic inputs'*

- 'waste quantity' the user defines the inflowing quantity of waste, design capacity and the total output for revenue.
- 'sorting facility', general economical properties about the sorting facility.
- 'transport vehicles'.
- 'revenue for recovered material'.

ATTENTION: The default values on total output for revenue are calculated based on the input waste flows. It is therefore strongly advised to use these default values (either fill out or leave blank), in order to avoid problems with non-closed material balances.

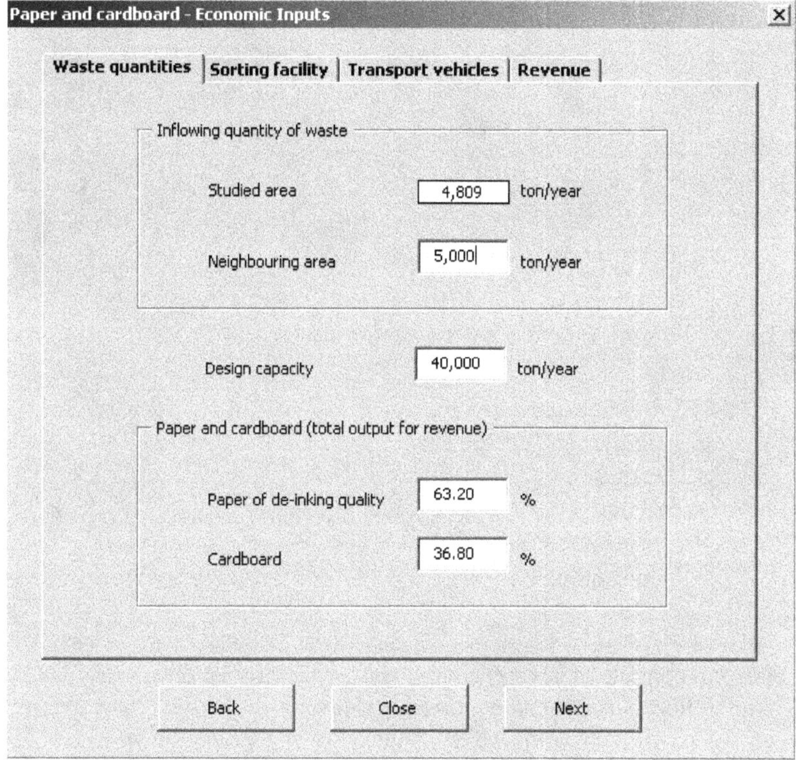

Figure 131. Waste recycling process – economic inputs

- *'social inputs'* (Figure 132)
  - visual impact, urban space, employment quality, and direct employment creation.

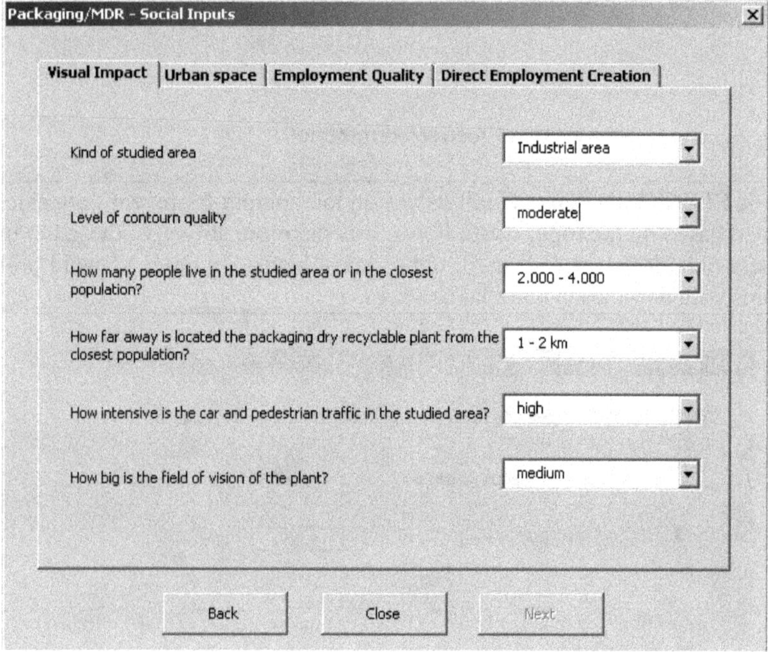

Figure 132. Waste recycling process - social inputs

Packaging/MDR recycling process can consist of up to 4 waste fractions

- plastics and composites
- metals
- glass
- paper & cardboard
- contamination

For each of these fractions a waste flow is calculated (e.g. plastics). The fraction consists of several components, which together make up 100% of the fraction. Since the contamination does not depend on the fraction, but on the whole Packaging/MDR stream, the contamination level is a % of the entire amount of input.

13.3.6.8 Secondary flows menu

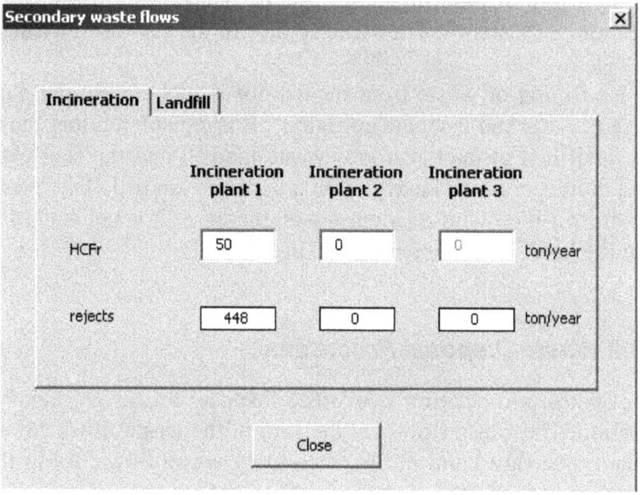

Figure 133. Secondary waste flow menu.

Treatment plants generate secondary waste flows, e.g. the waste outputs of MBP are low caloric fraction (LCFr) (which shows up in the secondary waste flow menu only if it is not utilised for 'land recultivation') and high caloric fraction (HCFr) (which shows up in the secondary waste flow menu only if it is not used for energy recovery in a cement kiln). Secondary waste flows from other plants are rejects. In each treatment plant module the user assigns the destination of the rejects, either to an incineration plant or to a landfill.

In the menu 'Secondary waste flows menu' the user sees the total amount of HCFr assigned for incineration. The tool allows the user to define to which incineration plant: 1, 2 or 3 these waste should be directed.

The amount of all rejects from recycling, composting and digestion which are going to the incineration plants is provided automatically by the tool. The rejects amount assigned to each incineration plant is shown in the 'secondary waste flows'.

Similarly, in the sheet 'landfill' the user assigns the total amount of the LCFr from MBP for landfilling to Landfill 1, 2 or 3. And he will see the waste rejects amount going to each landfill (the allocation follows automatically).

The Incineration plant modules will be deactivated if all inputs to the Incineration plant are =0. The same applies to the Landfill modules.

The landfilling of waste from incinerator is included in the incineration model (e.g. costs and leachate emission). It is considered that the fly ash is always landfilled at the hazardous waste landfill and the slag can be used for road construction or landfilled in a separate landfill. This waste is normally not landfilled with residual waste, therefore it is not considered as an input to the landfill model developed in this tool.

13.3.6.9 Waste Disposal Processes

The tool takes into account two waste disposal processes; landfilling and incineration. The waste flows are the sum of the waste flows defined in the 'primary waste flow' and in the 'secondary waste flow'. As in the 'waste treatment processes', the tool allows the user to define from one to three plants. The user defines the kind of treatment plant, the number of plants and the amount of waste treated in the primary flows menu and in the secondary waste menu.

13.3.6.10 General Treatment plant menu

Risk perception

This dialog menu defines the social indicator risk perception, which quantifies the population risk perception towards waste treatment plants by means of a public questionnaire survey.

The questionnaire survey procedure is described in detail in the Deliverable 5.

In the folder 'modules' there is a file called 'Questionnaire survey.xls'. This file has 4 sheets;

- *'Survey'*; this sheet presents the total questionnaires survey.

- *'Matrix'*; the user should introduce in this sheet the results of the survey. The user introduces the values result in the tool.

- *'Quantification'*; this sheet quantifies the answers introduced in the 'matrix sheet'.

- *'Exported values'*; the final results of the questionnaire survey are shown. These final results are used in the internal indicator calculation.

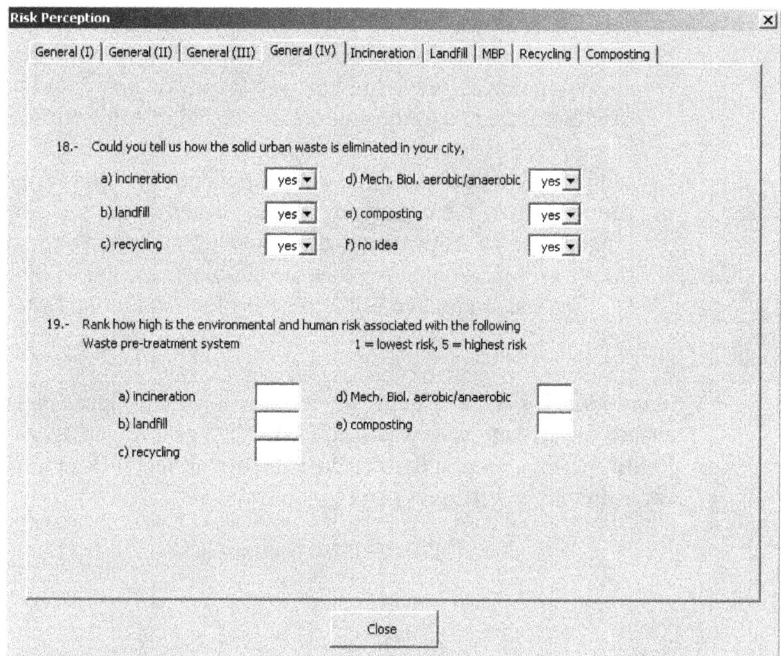

Figure 134. Risk perception indicator – social inputs

Compost credit

The dialog 'compost credit' assesses the application of produced compost (as well as sludge from the WWTP) on agricultural land. In this way the nutrients contained in the compost substitute artificial fertilisers (see Figure 135). Users not specialised in compost application can use the available default values in the compost credits section and leave the input field blank.

This dialog is divided into two sections:

- *'emissions to air'*;
    - soil application of peat; the avoided consumption of peat (by the introduction of organic carbon) is determined.
    - soil application of fertilisers; during the application of artificial fertilisers to agricultural soil a certain amount of NO_2 and ammonia is released. This section evaluates the emission factor related to the amount of N available in fertilisers.
    - sequestration potential; due to the application of compost to soil a share of the organic carbon is fixated in the soil for a long period (> 100 years); the so-called carbon sequestration (this is accounted for as a 'negative production of fossil CO_2'; it has a positive effect on the global warming potential).
- *'emissions to soil'*; The application of compost introduces nutrients to agricultural soil, which can substitute primary artificial fertilisers. This section defines the substitution factor (kg nutrient available in fertilisers per kg nutrient in compost)

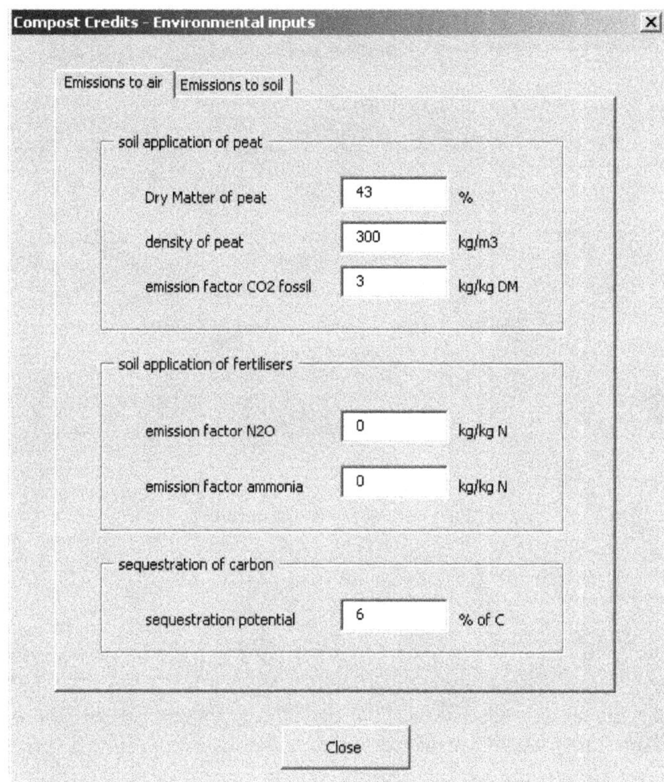

Figure 135. Compost credit – environmental inputs

Final destination

This dialog menu evaluates the social indicator; Final destination. It measures the social function of the used waste management option taking into account the recovery rate.

The user defines if the country is a new EU member state (accessed the EU in the year 2004) or not. The other variables are results from internal calculations. They are fixed and the user cannot modify them in this menu.

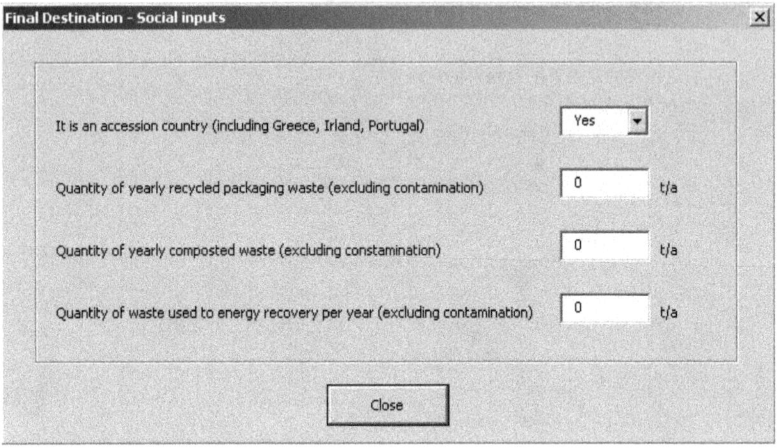

Figure 136. Final destination indicator – social inputs

Energy

Electricity

The tool takes into account the electricity balance of all the MSWMS processes considered (waste treatment, disposal, recycling & primary material manufacturing) transforming it into environmental loads (or LCI data) by considering the corresponding and country specific electricity supply mix.

If the tool electricity for the user country is not available, the user has to select in the tool a country with a similar 'electricity production mix' and with similar 'technologies'.

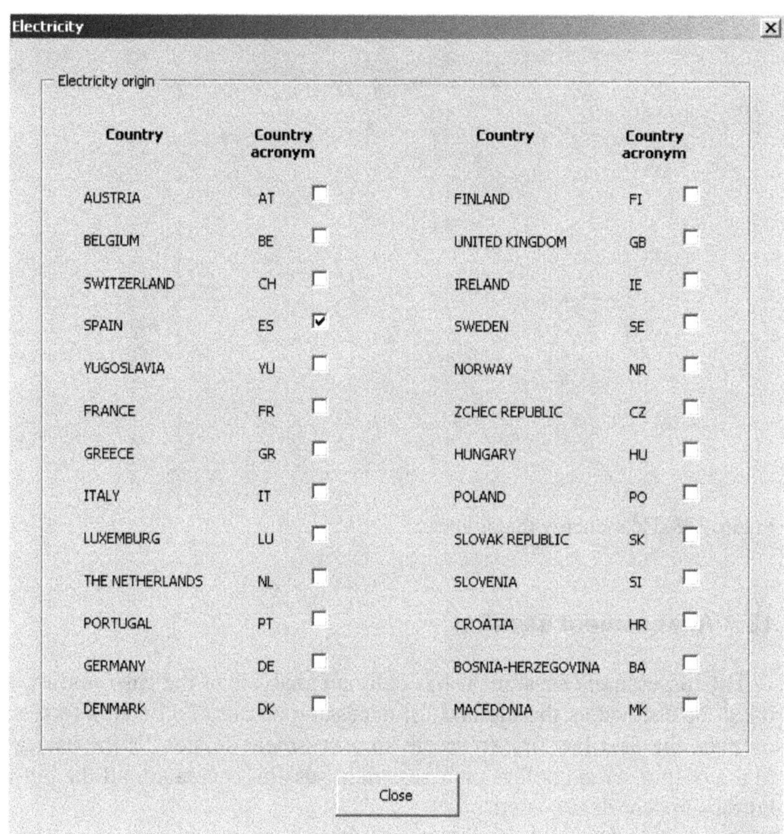

Figure 137. Electricity-energy dialog menu

Heat

The LCA-IWM tool considers the production and consumption of heat based on the origin of the energy supplier. The user indicates from which energy sources heat is produced in the country.

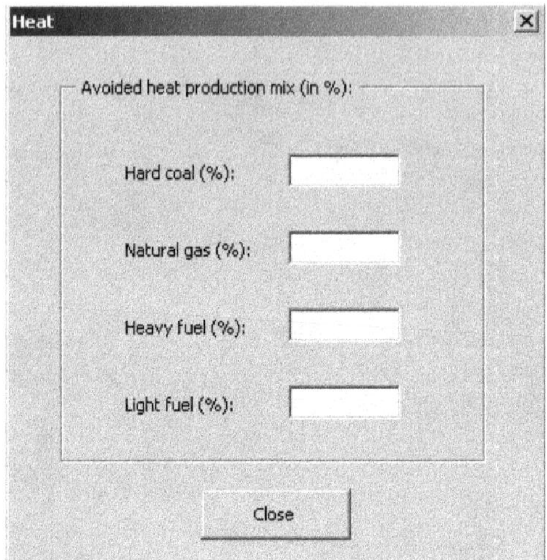

Figure 138. Heat-energy dialog menu

13.4 Assessment session

The 'assessment session' is basically an analysis of the case studies, and it can be defined as the method for accessing results. To be able to see the outputs, the user has first to create an assessment session. If the user goes to the output screen before an assessment session is created, all the outputs buttons appear deactivated.

When the user chooses: 'create a new assessment session':

- the user locates and names the assessment session. The session has to be located in the folder assessment session which is created when the user installs the tool.
- the user defines the scenarios which take part of the assessment session.

The user could create as many scenarios as he wishes; however, he could only assess a maximum of four scenarios and as a minimum one at a time on each assessment session.

After creating an 'assessment session' it is loaded, the user sees the files loading progress and the percentage of files loaded. This process will be longer or shorter depending on the number and complexity of the scenarios assessed.

If the user returns to the 'scenario mode' and modifies the system, the user should save the scenario. If the user wants to check the impact of the modifications in the outcomes, the 'assessment session' has to be loaded again.

When the user modifies something in a scenario, the user has always to update it (saving it). If the user wishes to update his session, it has to be loaded again. If the user does not load the session again, his modifications in the scenario are not considered in the output results.

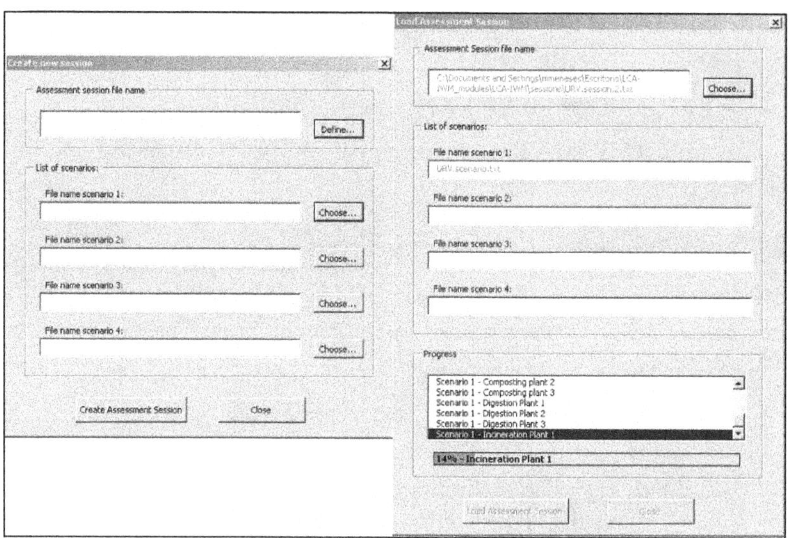

Figure 139. Assessment session dialog.

Assessment session options;

- *'New session'*; the user defines the name of the new assessment session, and the scenarios which compose this session. After creating a new session, the tool automatically loads this session.

- *'Load session'*; the user gets the session which was defined before. Once the user defines the session, the assessment results can be checked.
- *'Save session'*; the user saves the session, the properties of the session.

13.5 Assessment session results

The assessment session is divided into four sections: environmental, economic, social and condensed results. To switch one of these sections the user should select the tab pertaining to it. Almost all the assessment results are shown as tables and as graphics. In the output file in the output folder with the same name as the assessment session these tables and graphs are available as well in an Excel file for purposes of further results presentation.

The aim of the assessment session is to compare the scenarios impacts, in order to identify both the most negative and positive scenario and impacts.

13.5.1 Environmental assessment session

This section shows the environmental impacts evaluated by the tool; more in depth explanations are provided in Deliverable 3 and in section 4.2 of this Handbook.

The environmental assessment section could be divided into two main sections:

- Section 1; The environmental results shown are based on the LCA results. This section analyses six environmental impacts result for the waste management systems,
  - environmental impacts,
  - weighting factors,
  - weighted impact,
  - environmental impact for stages,
  - treatment plants.

- Section 2; The environmental results shown in this section are based on the specific targets of European waste policy. It could

be divided into packaging recovery and recycling targets and the diversion of biodegradable waste fraction from landfilling:
- packaging directive,
- biodegradable waste.

13.5.1.1 Environmental impacts

A brief description of the LCA-based criteria follows CML 2001 method used in the tool, more details in the section 4.2.:

- *'Depletion of abiotic resources';* depletion of natural resources taking into account the size of reserve for that resource in the ground and consumption rate.

- *'Global warming';* the impact of human emissions on the radio-active forcing (i.e. heat radiation absorption) of the atmosphere.

- *'Human toxicity';* the impact of toxic substances emitted to the environmental on human health.

- *'Photo-oxidant formation';* formation of reactive chemical compounds such as ozone by the action of sunlight on certain primary air pollutants.

- *'Acidification';* impact of acidifying pollutants on soil, ground-water, surface waters, living organisms and built environment.

- *'Eutrophication';* impacts of excessively high environmental levels for macronutrients, the most important of which are nitrogen and phosphorus.

In this section the six environmental impact based on LCA are shown for the total waste management system of the four scenarios. The graph and the table represent the total environmental impact value for each scenario.

A negative impact means an environmental benefit. A positive impact means an environmental burden (detriment).

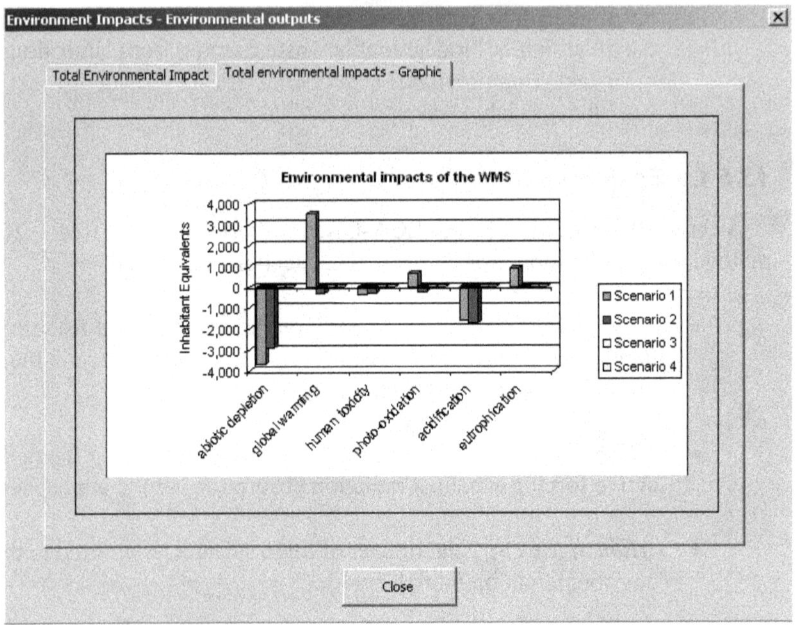

Figure 140. Environmental impacts

13.5.1.2 Environmental impact for stages

This section assesses the environmental impact for stages (Figure 141). The dialog box shows for each scenario the environmental impact of the three MSWMS stages (Temporary Storage, Collection & Transport and Treatment). This section allows the user to identify the impacts which contribute more and less to the MSWMS environmental impact within each scenario.

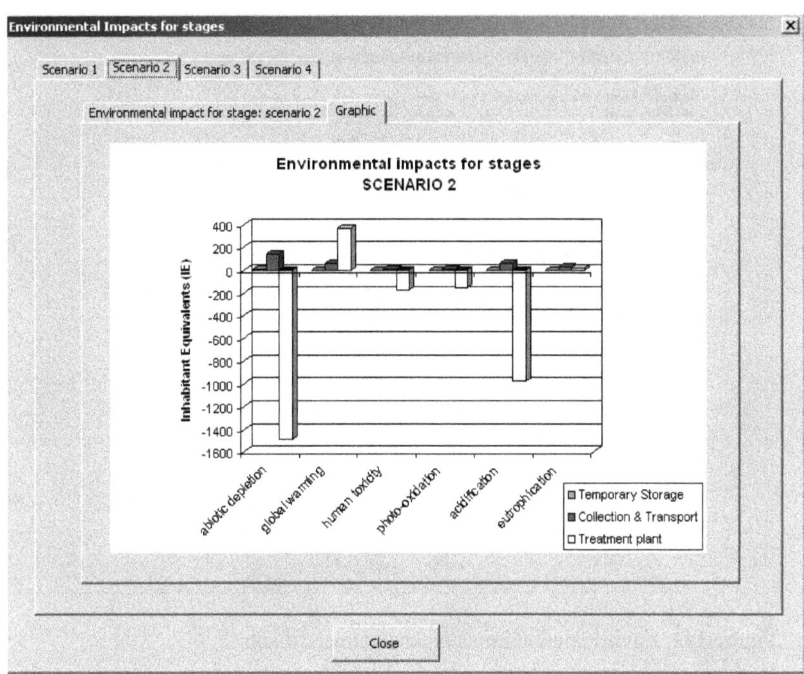

Figure 141. Environmental impact for stages.

13.5.1.3 Treatment plants

The LCA-IWM tool allows evaluating the environmental impacts for the single treatment plants. This section shows the treatment plant impact for each scenario independently.

These graphs are dynamic, and show the treatment plants which are considered in each scenario.

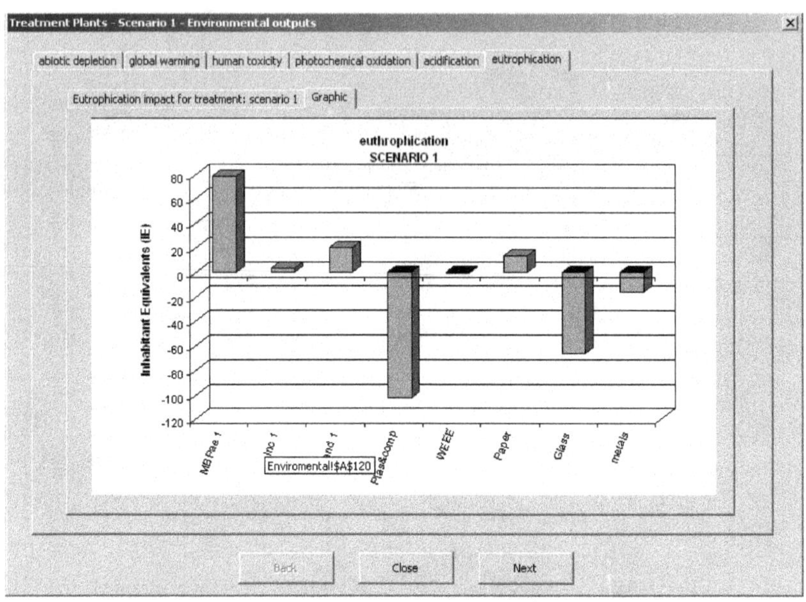

Figure 142. Environmental impacts for treatment plants.

13.5.1.4 Weighting factors/ weighted impacts

The impacts in the different categories can, in principle, not be simply added up. The German Environmental Agency suggests differentiation of importance of impacts categories. This differentiation is based on the endangering of the environment and the distance to environmental targets. With the weighting factors it is possible to estimate the aggregated environmental impacts.

The user can assign weighting factors to the environmental impacts. It is the responsibility of the user expert to choose the weighting factors which will be applied to the indicators (Figure 143). The weighting factor range is from (3 very important) to 0 (not important).

The default weighting factors values are:

- abiotic depletion and photochemical oxidant formation = 1
- acidification, eutropication and human toxicity = 2
- global warming = 3

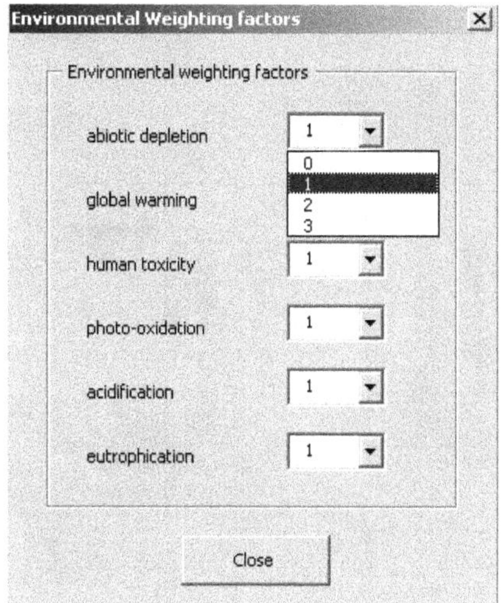

Figure 143. Environmental weighting factors.

When the user selects relative weighting factors, the application automatically updates the environmental weighted impacts. After that the user could assess the environmental weighted impacts (Figure 144). The dialog box is divided into two windows;

- *'Environmental weighted impact';* the total value of the environmental impact after applying the environmental weighting factors. These values have no physical meaning, but are merely given to show the contribution of each of the six indicators to the total condensed results.

- *'Graphic';* the environmental weighting (the user can choose the aesthetic outcome in graph form or table).

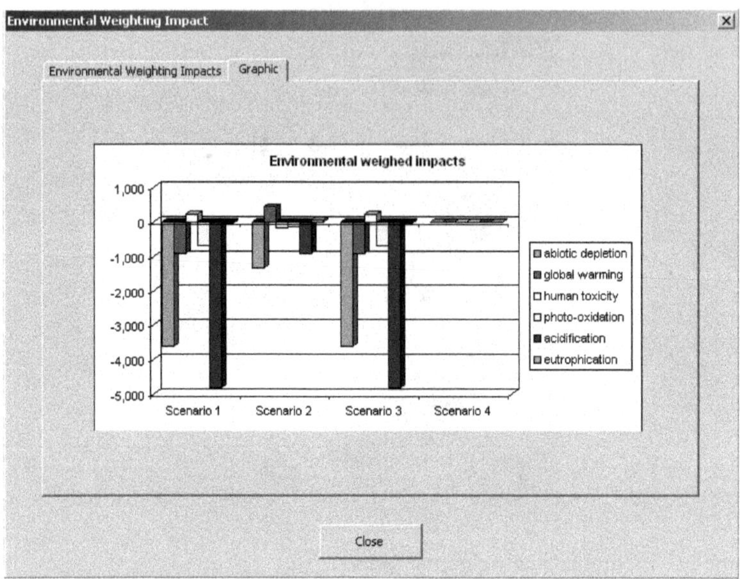

Figure 144. Environmental weighting impact.

13.5.1.5 Packaging directive

This section is based on the indicator recycling and recovery rates, which is based on specific EU waste policy targets. According to the Directive, recycling of packaging waste means 'the reprocessing in a production process of the waste materials for the original purpose or for other purposes including organic recycling but excluding energy recovery. The difference between recycling and recovery is that recovery also includes energy recovery.

The targets amount are based on the amounts of packaging waste generated by households and the amounts of packaging waste entering recycling or recovery facilities after sorting.

In this section the recycling and recovery rates of a scenario are compared to the valid targets for the year, material and country concerned (see Figure 145).

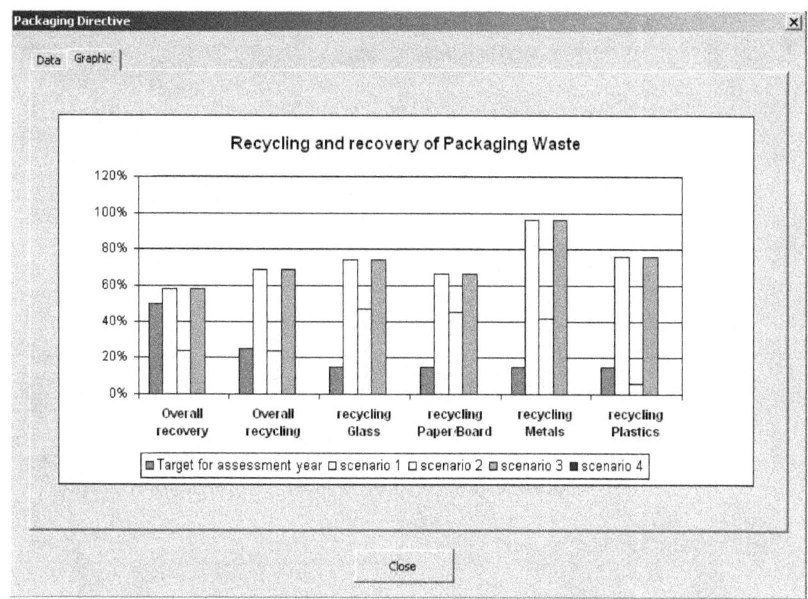

Figure 145. Recycling and recovery packaging waste.

13.5.1.6 Biodegradable waste

The Directive on landfill of waste (European Council 99/31/EC) aims at reducing the amount of biodegradable municipal waste going to landfill by the year 2016. The aim is to reach a reduction in the landfill of biodegradable waste to 35% of the total weight produced in 1995.

Limits for landfilling of biodegradable municipal waste, based on the EU landfill directive (99/31/EC):

- by 16.07.2006: 75% of the quantity produced in 1995
- by 16.07.2009: 50% of the quantity produced in 1995
- by 16.07.2016: 35% of the quantity produced in 1995

This section shows the results from the indicator reduction of biodegradable waste landfilled, based on specific EU waste policy targets. This indicator assesses whether a scenario ensures compliance with the targets of Landfill directive at a city level and the household waste (see Figure 146).

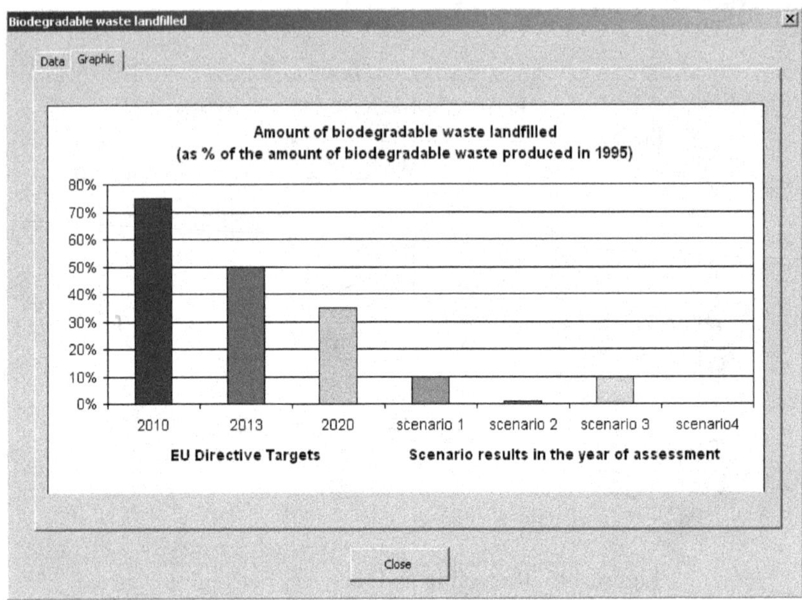

Figure 146. Amount of biodegradable waste landfilled.

13.5.2 Economic assessment session

This section describes the economic impacts evaluated by the tool; more in depth explanations are provided in Deliverable 4 and in section 5.2 of this Handbook.

The economic assessment section are divided into two main sections:

- *Section 1;* the economic impacts results shown are for the total impact and for the subsystem efficiency.

- *Section 2;* efficiency at the municipality level, equity and dependence on subsides.

13.5.2.1 Economic impacts

The indicators studied are the Annual total cost and the Annual total revenue. These impacts are presented for all the waste management system in a table and a graph (Figure 147).

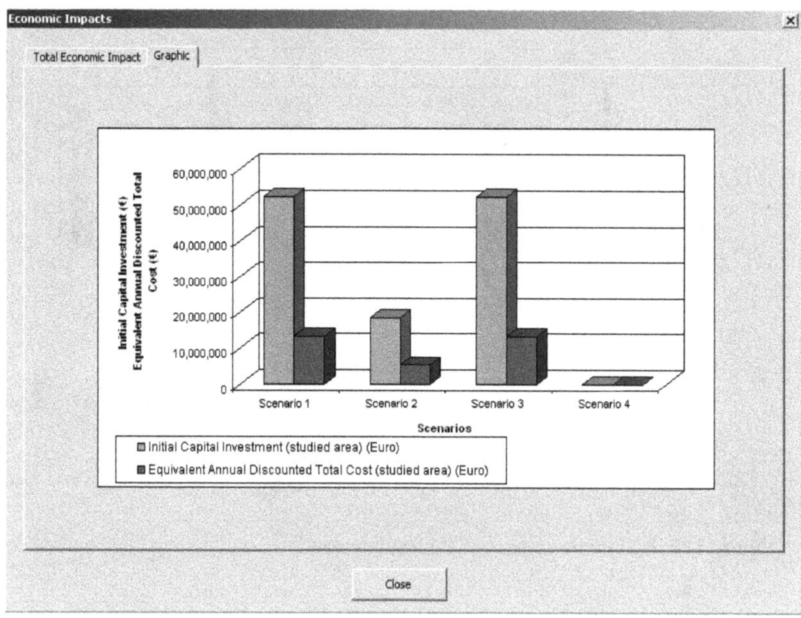

Figure 147. Economic impacts for the waste management system

13.5.2.2 Subsystems impacts

In this section the Annual total cost of MSWMS per ton of waste, household and waste are shown for each subsystems of the MSWMS, and for each scenario independently (Figure 148).

This assessment allows the user to know the waste management stage with the highest as well as the lowest economic impact.

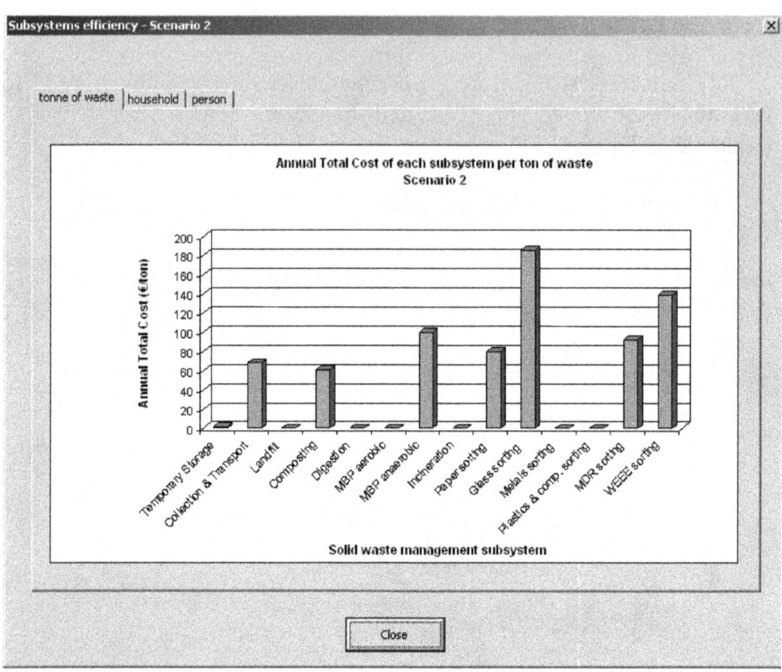

Figure 148. Economic impact of each subsystem.

13.5.2.3 Efficiency, equity, subsidies

This section presents the results of the Efficiency at a municipality level, Equity and the Dependence on subsidies. The Efficiency evaluates the impact of the total system at a municipality level. The Equity evaluates the extent to which the economic burden is distributed equitably among neighbourhoods and the public. The aspect Dependence on subsides examines the extent to which the municipality is relying on 'external' sources, i.e. on grants and subsidies (Figure 149).

These dialogue is divided into three menus:

- *'Efficiency at a municipality'*, this economic impact is measured by means of six aspects:

- Annual total cost of MSWMS per ton of waste (euro/ton)
- Annual total cost of MSWMS per household (euro/hh)
- Annual total cost of MSWMS per person (euro/person)
- Revenue from recovered material & energy
- Municipal solid waste management cost as % of GNP of the city (%)
- Diversion between revenue and expenditure for MSWMS

- *'Equity'*, the economic equity of the waste management system is determined by the next two aspects:

  - Cost per person as % of minimum wage (%)
  - Cost per person / income per person (%)

- *'Subsidies' (dependence of subsidies)*, :

  - Subsidies or grants per person (euro/person)

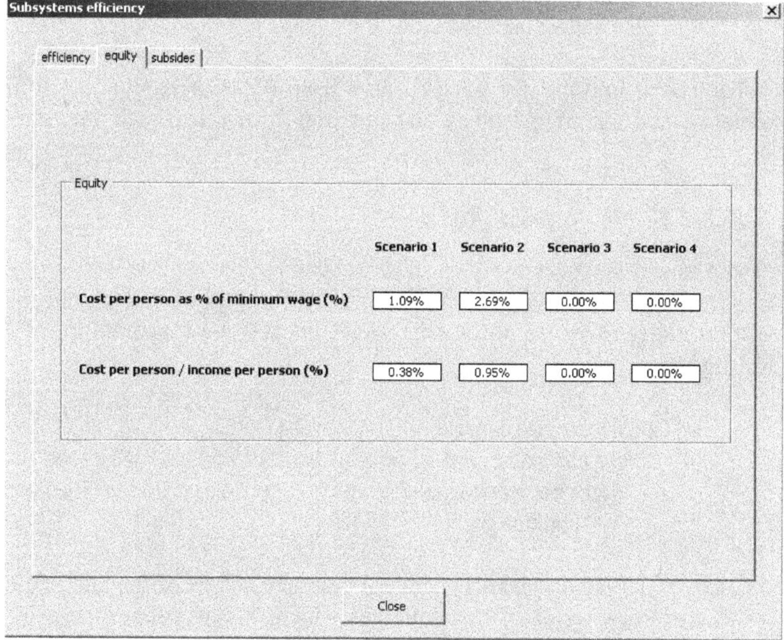

Figure 149. Subsystem efficiency of the MSWMS.

13.5.3 Social assessment session

This section quantifies the social sustainability of the waste management system; more in depth explanations are provided in the handbook and in Deliverable 5.

The social impacts evaluated by the tool are:

- Noise
- Odour
- Visual Impact
- Employment Quality
- Employment Creation
- Private Space
- Public Space
- Risk perception
- Complexity
- Comfort
- Final Destination
- Distribution and location

The social impacts can have a value from 0 (the best situation from a social point of view) to 1 (the worst situation from a social point of view).

13.5.3.1 Social impact

The social impacts are classified depending on the MSWMS stages. Hence, not all the social impacts are relevant for all the waste management system stages. Because of this, the social impacts are classified in the following way:

- Temporary Storage
  - *'Odour'*; potential of odour nuisance caused by a given waste temporary storage system to the public, it undermines social acceptance for a MSWMS.

  - *'Visual impact'*; change in the appearance of the landscape as a result of development which in this paper's case has a negative intrinsic visual impact.

- *'Convenience';* the distance the public has to travel to be able to dispose of his waste, it determines the social acceptable for a MSWMS.

- *'Urban space'*; the potential space requirement for waste temporary storage.

- *'Private space';* a quantitative assessment of the private space consumption acceptability required by the waste temporary storage.

- *'Noise';* it is expressed in average container filling noise potential.

- *'Complexity';* the acceptability of the understanding of temporary storage.

- *'Distribution and location';* the equity distribution and location of temporary storage.

- Collection & Transport
  - *'Traffic';* traffic mileage represents the total distance driven by all collection and transport vehicles of a certain waste fraction per year.

  - *'Urban space'*; the potential space requirement for waste collection and transport.

  - *'Noise';* is expressed through the increase of the mean sound level per year caused by the additional traffic through the collection of waste in relation to the existing background traffic sound level.

  - *'Employment quality';* assesses the job characteristic and the work and the wider labour market context.

  - *'Employment quantity';* is the total amount of direct employment in a MSWMS due to the collection and transport stage.

  - *'Block potential';* traffic blocking potential expresses the total time all collection vehicles spend per year during collection.

- Treatment plants
  - *'Odour';* describes potential of odour nuisance caused by a given waste management installation to the nearby public.

  - *'Visual impact';* takes into account the visibility, fragility and contour quality.

  - *'Traffic';* the traffic mileage expresses the nuisance which is caused by the extra traffic caused by a MSWMS, by extra driven kilometers.

  - *'Urban space';* is the potential land consumption by waste treatment infrastructure, it considers the total area occupied and the kind of occupation.

  - *'Employment quality';* assesses the job characteristic and the work and the wider labour market context in the waste treatment plants.

  - *'Employment quantity';* evaluates how much employment is created due to the waste treatment stage of the MSWMS in a city.

  - *'Risk perception';* quantifies the population risk perception towards a waste treatment plant by means of a questionnaire survey of the population

  - *'Final destination';* the social function of the used waste management system option taking into account the recovery rate.

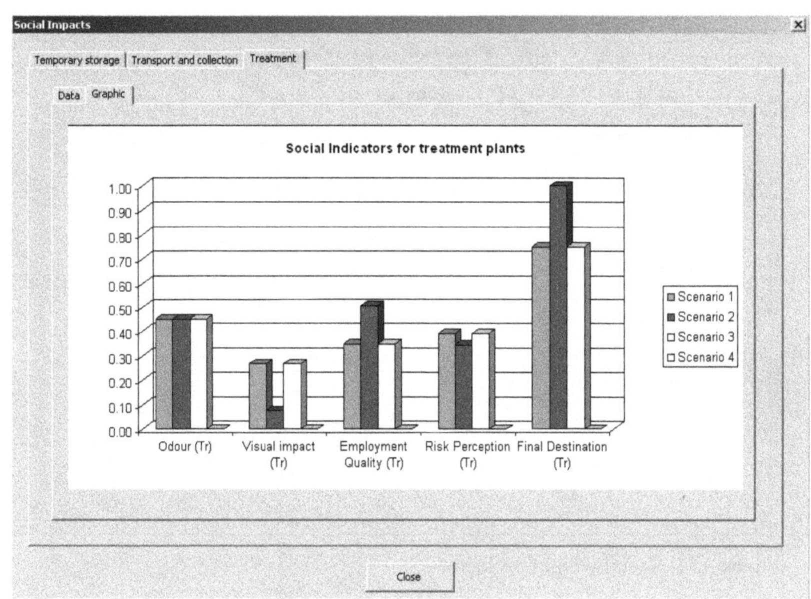

Figure 150. Social impact for stages.

13.5.3.2 Social weighting factors and social weighting impacts

The tool allows the user to assign weighting factors to the social impacts. The weighting factor range is from 0 (irrelevant impact) to 3 (relevant impact) (Figure 151). Depending on the weighting factors assigned, the user automatically receives weighted social impacts.

The social weighted impacts are shown in three groups; Temporary Storage, Collection & Transport and Treatment. These results are shown in a table as well as in a graph. These values in themselves have no physical meaning, but are merely given to show the contribution of each of the indicators to the total condensed results.

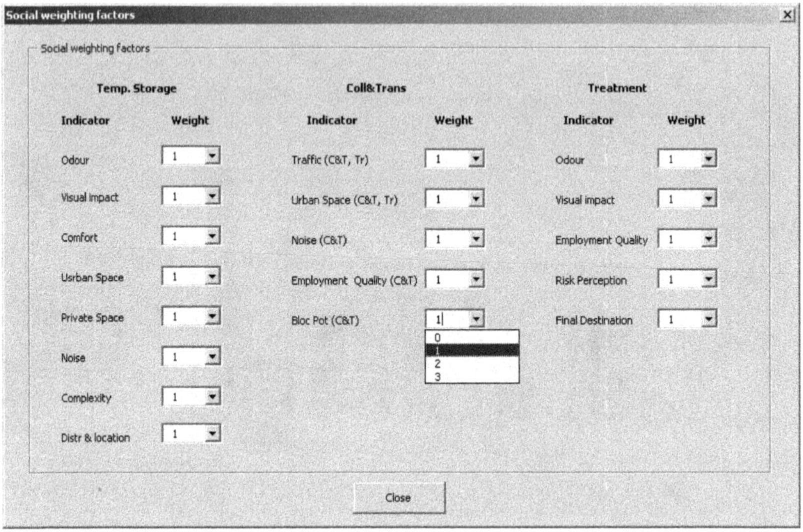

Figure 151. Social weighting factors.

13.5.3.3 Social impacts overview

All the social impacts are displayed in a radar chart. This menu shows in a graph (Figure 152) all the social impacts considered for all the studied scenarios. In this way the user assesses the most and less relevant social aspect. In the output file the results are presented in another graph type as well.

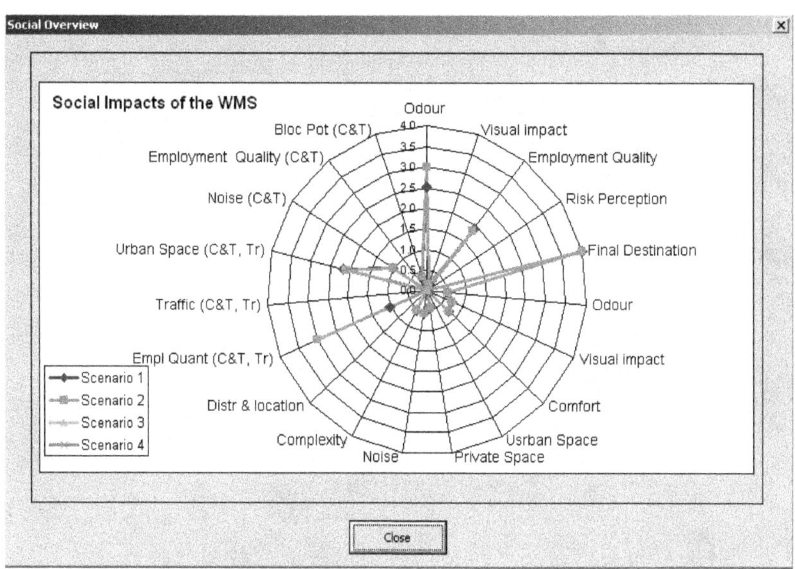

Figure 152. Social overview graphic.

13.5.4 Condensed results

This section shows a global summary for the three sustainability aspects studied, environmental, economic and social, for all the scenarios assessed.

13.5.4.1 Relative global impact

This section presents the relative importance of the different scenario. The user can have a quick overlook of the scenario with highest impact, and also about the sustainability aspect with both, worst and best impact (Figure 153).

This relative global impact evaluates the impact of a scenario relative to scenario 1. Because of this, the relative global impact for scenario 1 is equal to 100. In case of the environmental impacts this value may also be -100, which represents an environmental gain. For further details on the specific properties of the environmental indicators (they can be positive or negative) and their aggregation, please refer to section 4.4. Economic and social indicators only show positive results.

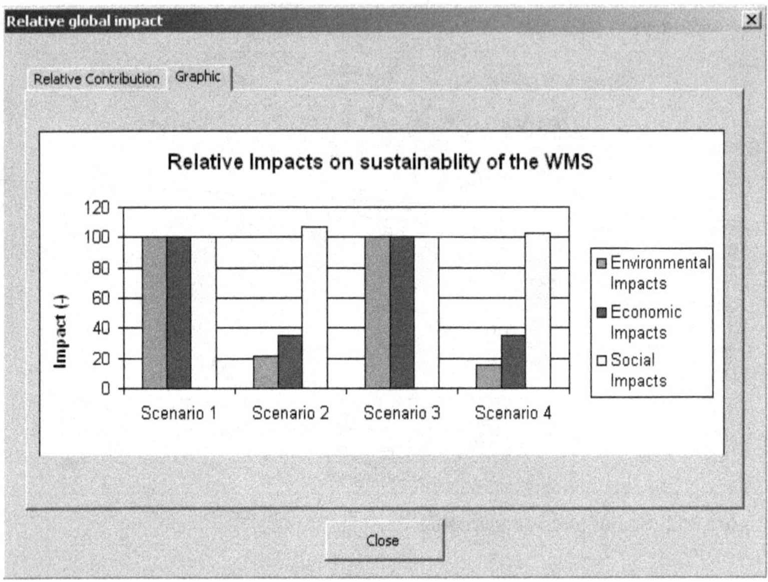

Figure 153. Relative global impact of the threes aspects of sustainability

13.5.4.2 Compliance with EU waste policy targets

This menu summarises the compliance with the packaging and landfill directive by means of a 'traffic light'. The color depends on the scenario result;

- Recovery:
  - Red = target is not met.
  - Green = target is met
  - White = unavailable data
- Recycling:
  - Red = general target is not met.
  - Orange = general target is met, at least one material specific target is not met.
  - Green = general and material targets are met
  - White = unavailable data
- Landfilling of biodegradable waste:

Red = target is not met.
Green = target is met
White = unavailable data

ATTENTION: Recovery/recycling: the targets are met in the case where the results exceed the targets.
Landfilling of biodegradable waste: the target is met in the case where the results are below the target

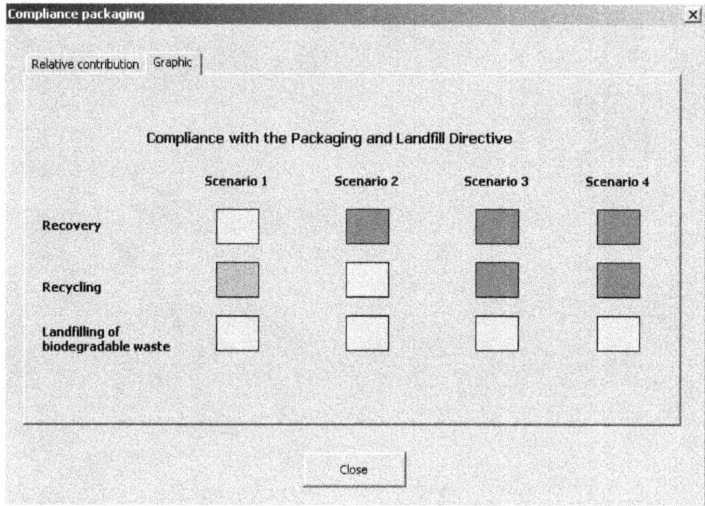

Figure 154. Compliance with the packaging and landfill directive

13.6 Remarks

This tutorial was intended to give the user an overview of the capabilities of the LCA-IWM interface. The LCA-IWM Project Partners wish to deliver software and a database with the fewest possible bugs but this goal cannot be attained in one day. The existing release is open to improvements. The Consortium would be pleased to receive any detailed description of what is considered a problem or weakness by specifying the release used, the hardware configuration, the context in which the problem occurs and any other relevant information (contact via www.lca-iwm.net).

References

Armstrong, J.S., Principles of forecasting: a handbook for researchers and practitioners, Kluwer Academic Publishers, Boston, 2001

Bach, H., Mild, A., Natter, M. and Weber, A. (2004) Combining socio-demographic and logistic factors to explain the generation and collection of waste paper, Resources, Conservation and Recycling, Volume 41, Issue 1, pp. 65-73

Balzari, Schudel, Grolimund and Petermann (1991). Straßenlärmmodell für überbaute Gebiete. Schriftenreihe Umwelt Nr. 15. Bundesamt für Umwelt, Wald und Landschaft, 3. Auflage, Bern

Bilitewski Bernd, Härdtle Georg, Marek Klaus (1996) Waste Management. Springer

Bockreis, A., Steinberg, I., Rohde, C. and Jager, J. (2003) Gaseous emissions of mechanically-biologically pre-treated waste from long-term experiments, in : Proceedings of the Ninth International Waste Management and Landfill Symposium 6-10 October 2003, S. Margherita Di Pula (Cagliari), Sardinia

Büchl, R. and Loll, L (2002) Optimierung der Herstellung von Ersatzbrennstoffen, Studienarbeit im Institut WAR, Technische Universität Darmstadt, Darmstadt

BZL - Kommunikation und Projektsteuerung GmbH and DPU – Deutsche Projekt Union (2000) Anforderungen an Ersatzbrennstoffe aus Abfällen für die Zementindustrie, Auftraggeber: Ministerium für Umwelt und Verkehr des Landes Baden-Württember, Oyten

CEC, Commission for the European Communities, (2000) Social Policy Agenda, Brussels C5-0508/2000-2000/2219 (COS).

Cerbe & Hoffmann (1994) Einführung in die Thermodynamic; 10 Auflage, Hansen Veralg, München

Conesa, V. (2003), Guía Metodologica para la evaluación del impacto ambiental, Ediciones Mundi-Prensa, Madrid.

Dach, J. (1998) Zur Deponiegas- und Temperaturentwicklung in Deponien mit Siedlungsabfällen nach mechanisch-biologischer Abfallbehandlung, Dissertation am Fachbereich 13 Bauingenieurwesen der Technischen Universität Darmstadt, Darmstadt

Dehoust, G., Gebhardt, P. and Gärtner, S. (2002) der Beitrag der thermischen Abfallbehandlung zu Klimaschutz, Luftreinhaltung und Ressourcenschonung; in Öko-Institut e.V, Darmstadt

Dennison, G.J., V.A. Dodd, and B. Whelan, A socio-economic based survey of household waste characteristics in the city of Dublin, Ireland, II. Waste quantities. Resources, Conservation and Recycling, 17, 245-257, 1996.

EC, European Commission (2001), Employment and social policies: A framework for investing in quality, Communication from the Commission to the Council, the European Parliament, the Economic and Social Committee and the Committee of the Regions, COM 313 final 20.06.2001

Ecobalance UK and DMG Consulting Ltd (1999) Life cycle assessment and life cycle financial analysis of the proposal for a directive on waste from electrical and electronic equipment. Final report, July 1999 (for the Department of Trade and Industry of the United Kingdom).

Ecobilan (2003) – WM3.4rl: WISARD Reference guide, Ecobilan – WM3.4rl

ECOINVENT-2000 (2003) Centre for LCIs in the ETH domain; LCI of electricity supply mix in European countries.

European Commission Decision 2000/532/EC replacing Decision 94/3/EC establishing a list of wastes (European Waste Catalogue) Consolidated TEXT produced by the CONSLEG system of the Office for Official Publications of the European Communities; CONSLEG: 2000D0532 — 01/01/2002, available at: http://europa.eu.int/eurlex/en/consleg/main/2000/en_2000D0532_index.html

European Environment Agency (EEA) Glossary, available at: http://glossary.eea.eu.int/EEAGlossary/searchGlossary

European Environment Agency EEA (2002) Biodegradable municipal waste management in Europe, Part 1 Strategies and instruments, Topic report 15/2001

European Environment Agency EEA (2003a) EEA multilingual environmental glossary, available at: http://glossary.eea.eu.int/EEAGlossary/H/household_waste

European Parliament (2000) Directive 2000/76/EC of the European Parliament and of the Council of 4 December 2000 on the incineration of waste, Official Journal L 332 , 28/12/2000 P. 0091 - 0111 available at: http://europa.eu.int/eur-lex/en/lif/reg/en_register_15103030.html

European Parliament (2002) Directive 2002/96/EC of the European Parliament and of the Council of 27 January 2003 on waste electrical and electronic equipment (WEEE) - Official Journal L 037 , 13/02/2003 P. 0024 – 0039, available at: http://europa.eu.int/eur-lex/en/lif/reg/en_register_15103030.html

European Parliament and Council (1994) Directive 94/62/EC of 20 December 1994 on packaging and packaging waste. Official Journal L 365, 31/12/1994 P. 0010-0023, available at: http://europa.eu.int/eur-lex/en/lif/reg/en_register_15103030.html

European Parliament and Council (2004) Directive 2004/12/EC of 11 February 2004 amending Directive 94/62/EC on packaging and packaging waste. Official Journal L 47, 18/02/2004 P. 0026-0031. available at: http://europa.eu.int/eurlex/en/lif/reg/en_register_15103030.html

European Topic Centre on Waste and Material Flows (ETC-WMF) (2003) Preparing a waste management plan. A methodological guidance note, available at: http://waste.eionet.eu.int/Publications

European Union (EU) (2004) Economic Forecasts, Spring 2004.

FAO (Food and Agriculture Organization of the UN) (2004) http://apps.fao.org/lim500/wrap.pl?Population.LTI&Domain=SUA&Language=english, February 2004.

Fischer, C.; Crowe, M.: "Household and municipal waste: Comparability of data in EEA member countries". European Environment Agency, topic report No 3/2000, Copenhagen 2000.

Flamme, S. (2002) Energetische Verwertung von Sekundärbrennstoffen in industriellen Anlagen – Ableitung von Maßnahmen zur umweltverträglichen Verwertung, Dissertation vom Fachbereich Bauingenieurwesen der Bergischen Universität – Gesamthochschule Wuppertal, Wuppertal, available at: http://elpub.bib.uniwuppertal.de/edocs/dokumente/fb11/diss2002/flamme;internal&action=buildframes.action

FNR (2004) Trockenfermentation –Evaluierung des Forschungs und Entwicklungsbedarfs. Gülzower Fachgespräche: Band 23 - Fachagentur Nachwachsende Rohstoffe, Gülzow, Germany.

Fricke, K., Franke, H., Dichtl, N., Schmelz, K.-G., Weiland, P. and Bidlingmaier, W. (2002c) Biologische Verfahren zur Bio- und Grünabfallverwertung. In: Loll, U. (Eds.), ATV Handbuch – Mechanische und biologische Verfahren der Abfallbehandlung. Ernst & Sohn Verlag für Architektur und technische Wissenschaften, GmbH, Berlin, Germany.

Fricke, K., Hake, J., Hüttner, A., Müller, W., Santen, H., Wallmann, R. and Turk, T. (2003) Aufbereitungstechnologien für Anlagen der mechanisch-biologischen Restabfallbehandlung, No: 5615; MuA Lfg. 4/03 in Müll Handbuch, Band 5; Erich Schmidt Verlag; Berlin

Fricke, K., Müller, W. (1999) Stabilisierung von Restmüll durch mechanisch-biologische Behandlung und Auswirkungen auf die Deponierung, Verbundsvorhaben „Mechanisch-Biologische Behandlung von zu deponierenden Abfällen", Witzenhausen

Fricke, K., Müller, W., Santen, H., Wallmann, R. and Ziehmann, G. (2002a) Stabilitätskriterien für biologisch behandelten Restmüll, Konsequenzen für den Betrieb von MBA-Anlagen und Deponien; No: 5614; MuA Lfg. 11/02 in Müll Handbuch, Band 5; Erich Schmidt Verlag; Berlin

Fricke, K., Niesar, M. and Turk, T. (2002b): Restabfallmengen und -qualitäten für die mechanisch-biologischen Restabfallbehandlungsverfahren; No: 5616; MuA Lfg. 11/02 in Müll Handbuch, Band 5; Erich Schmidt Verlag; Berlin

Goodland, R. (2002) Sustainability: Human, Social, Economic and Environmental; in: Encyclopedia of Global Environmental Change; John Wiley & Sons

Guinée J.B., Gorrée M., Heijungs R., Huppes G., Kleijn R., De Koning A., Van Oers L., Wegener Sleeswijk A., Suh S., Udo de Haes H.A., De Bruijn H., Huijbregts M.A.J., Lindeijer E., Roorda A.A.H., Van derVen B.L. and Weidema, B.P. (2001) Handbook on Life Cycle Assessment; operational guide to the ISO standards. Kluwer Academic Publishers, Dordrecht

Hellweg, S. 2000: Time- and Site- Dependent Life Cycle Assessment of Thermal waste Treatment Processes; Dissertation in Swiss Federal Institute of Technology, Zurich

Hellweg, S., Doka, G., Finnveden, G. and Hungerbühler, K. (2003) Ecology: Which Technologies Perform Best?, in Municipal Solid waste Management, Ludwig, Hellweg anf Stucki (eds) Springer-Verlag, Berlin-Heidelberg

Hellweg, S., Hofstetter, T.B. and Hungerbühler, K., (2001) Modelling waste incineration life cycle-inventory analysis in Switzerland, in Environmental Modelling and Assessment 6: Kluwer Academic Publishers (Eds.), Netherlands

Heyer, K.-U. and Stegmann, R. (2001) Leachate management: leachate generation, collection, treatment and costs, available at: http://home.t-online.de/home/Karsten.Hupe/pdf/leachate.pdf

Huijbregts, M. (1999) Life cycle impact assessment of acidifying and eutrophying air pollutants, Draft version, Interfaculty Department of Environmental Science, available at: http://www.leidenuniv.nl/interfac/cml/ssp/projects/lca2/report_mh_iias a2.pdf

Huijbregts, M.A.J. (2000) Calculation of toxicity potentials for ethylene oxide and hydrogen fluoride. Institute for Biodiversity en Ecosystem Dynamics, University of Amsterdam, Amsterdam.

ifu and ifeu (2001) Umberto, Software für das betriebliche Stoffstrommanagement. Institut für Umweltinformatik Hamburg GmbH and Institut für Energie- und Umweltforschung Heidelberg GmbH, Germany.

IKW and I&U (2000) Study on the Feasibility of Mechanical-biological Residual Waste Treatment Plant in Phitsanulok / Thailand. On behalf of Deutsche Gesellschaft für Technische Zusammenarbeit (GTZ) GmbH. By Beratungsinstitut für Komunalwirtschaft GmbH (IKW) and Infrastruktur & Umwelt Professor Böhm und Partner(I&U).

International Source Book on Environmentally Sound Technologies (ESTs) for Municipal Solid Waste Management, 1996 UNEP.Available at: http://www.unep.or.jp/ietc/ESTdir/Pub/MSW/SP/SP3/SP3_2.asp

Jenkins, R.R., Martinez, S.A., Palmer, K., Podolsky, M.J. (1999) The Determinants of Household Recycling: A Material Specific Analysis of unit Pricing and Recycling Program Attributes, Resources for the Future, available at: http://www.rtf.org

Johnke, B. (2003) Emissions from Waste Incineration, in: Good Practice Guidance and Uncertainty Management in National Greenhouse Gas Inventories, Institute for Global Environmental Strategies; available at: www.ipccnggip.iges.or.jp/public/gp/bgp/5_3_Waste_Incineration.pdf

Karavezyris, V., Prognose von Siedlungsabfällen: Untersuchungen zu determinierenden Faktoren und methodischen Ansätzen, TK Verlag, Neuruppin, 2001.

Kern, M. (1999) Stand und Perspektiven der biologischen Abfallbehandlung in Deutschland. In: Wiemer, K. and Kern, M., (Eds.). Bio- und Restabfallbehandlung III. M.I.C. Baeza-Verlag, Witzenhausen, Germany, pp. 293-321.

Kern, M. (2001) Mechanisch-biologische Restabfallbehandlungsanlagen in Deutschland MuA Lfg. 7/01 in Müll Handbuch, Band 5; Erich Schmidt Verlag; Berlin, Germany.

Krümpelbeck, I. (1999) Untersuchungen zur langfristigen Verhalten von Siedlungsabfalldeponien, Dissertation am Fachbereich Bauingenieurwesen der Bergischen Universität – Gesamthochschule Wuppertal

Krümpelbeck, I. and Ehrig, H.J. (2001) BMBF-Forschungsvorhaben: Abschätzung der Restemissionen von Deponien in der Betriebs- und Nachsorgephase auf der Basis realer Überwachungsdaten, No: 4587; MuA Lfg. 3/01 in Müll Handbuch, Band 4; Erich Schmidt Verlag; Berlin

Lahl, U. (2001) Entscheidungshilfen durch stoffstromanalytische Betrachtungen bei der Bewertung von abfallwirtschaftlichen Maßnahmen, Habilitationsschrift (unpublished), Technische Universität Darmstadt, Institut WAR, Darmstadt

Ludwig C. Hellweg, S. and Stucki S. (Eds) (2003) Municipal Solid Waste Management, Springer, Berlin, Heidelberg, Germany

Market and Opinion Research International (MORI), 1999 Aylesford Newsprint Recycling Report, http://www.wasteguide.org.uk/-issues/mn_public_research_public.stmf

McDougall, F., White, P., Franke, M. and Hindle, P. (2001) "Integrated Solid Waste Management: a Life Cycle Inventory", Blackwell Science Ltd, Oxford

McDougall, Forbes; White, Peter; Franke, Marina; Hindle, Peter (2001): Integrated Solid Waste Management. A Life Cycle Inventory. Blackwell Publishing.

Morf, L.S. und Brunner, P.H. (1999) Methoden zur indirekten Bestimmung der Zusammensetzung von Siedlungsabfällen; No 1755; MuA Lfg. 7/99 in Müll Handbuch, Band 3; Erich Schmidt Verlag; Berlin.

Müller, W., Wallmann, R., Hake, J. and Turk, T. (2001): Stand der Technik und Entwicklungspotenziale der mechanisch-biologischen Restabfallbehandlung, in Bio- und Restabfallbeh. V, Witzenhausen-Institut (eds), Witzenhausen

Müller-Wenk, R. (2002) Zurechnung von lärmbedingten Gesundheitsschäden auf den Straßenverkehr, Schriftenreihe Umwelt Nr, 339, Bundesamt für Umwelt, Wald und Landschaft, Bern

OECD (2002) National Accounts for OECD Countries, 1989-2000.

Öko-Institut e.V. (1999) Waste Prevention and Minimization (Final Report), Commissioned by the European Commission, DG IX.

Panagiotakopoulos D. (2002) Sustainable Municipal Solid Waste Management, Zygos, Thessaloniki, Greece (in Greek).

Poiesz Th.B.C. (1999) Gedragsmanagement; waarom mensen zich (niet) gedragen. Inmerc bv, Wormer

RDC-Env. & Pira Int. (2003) Evaluation of costs and benefits for the achievement of reuse and recycling targets for the different packaging materials in the frame of the packaging waste directive 94/62/EC, available at: http://europa.eu.int/comm/environment/waste/studies/

Rettenberger, G. and Schneider, R. (1996) Überblick über die Anforderungen und den Stand der Sickerwasserreinigungstechnik, in: Trierer Berichte zur Abfallwirtschaft, Band 10 - Wirtschaftliche Sickerwasserreinigung, Economica Verlag, Bonn, S. 9 – 43

Rettenberger, G. and Urban-Kiss, S (2000) Beispiel fur die UVP bei der Deponieplanung MuA Lfg. 6/00 in Müll Handbuch, Band 4; Erich Schmidt Verlag; Berlin

Robinson, H.D., Knox, K., van Santen, and Tempany, P.R. (2002) Compliance of UK Landfills with EU Pollution Emission Legislation: Development of a Reporting Protocol, available at: http://www.leachate.co.uk/Leachate-Downloads/Trace-Organics.doc

Rodrigo J. and Castells F. (2000) Environmental evaluation of different strategies for the management of municipal waste in the region of Catalonia (project founded by Junta de Residus, the regional Waste Agency of Catalonia).

Rotter, S. (2004) Schwermetalle in Haushaltsabfällen; No 2829; MuA Lfg. 1/04 in Müll Handbuch, Band 4; Erich Schmidt Verlag; Berlin.

Salhofer S., Grassinger D., Lebersorger S., Graggaber M.; Potential und Maßnahmen zur Vermeidung kommunaler Abfälle am Beispiel Wiens (Kurzbericht), Universität für Bodenkultur, Abteilung Abfallwirtschaft, Wien, Dezember 2000.

Schwing, E. (1999) „Bewertung der Emissionen der Kombination mechanisch-biologischer und thermischer Abfallbehandlungsverfahren in Südhessen", Dissertation, Institut WAR, TUD Darmstadt, WAR-Schriftenreihe Bd. 111, Darmstadt

Sircar, R., F. Ewert, and U. Bohn, Ganzheitliche Prognose von Siedlungsabfällen, Müll und Abfall, 1, 7-11, 2003.

Smith, A., Brown, K., Ogilvie, S., Rushton, K. and Bates, J. (2001) Waste management options and climate change. Report to the European Commission. Office for Official Publications of the European Communities, Luxembourg.

Soyez, K. et al. (2000) Gesamtdarstellung der Wissenschaftlichen Ergebnisse des Verbundsvorhabens, Verbundsvorhaben „Mechanische - Biologische Behandlung von zu deponierenden Abfällen", Potsdam

Soyez, K., Plickert, S. and Koller, M. (2001b) Von der Abfall- zur Rohstoff und Energiewirtschaft: Umsetzung der Ziele Nachhaltigkeit, Klimaschutz und Ressourceneffizienz in der MBA-Technologie, in 4 Wetzlarer Abfalltagung, available at: http://www.gts-oekotech.de/docs/Beitrag_Wetzlarer_Abfalltage_2001.PDF

Soyez, K., Thrän, D., Hermann, T., Koller, M. and Plickert, S. (2001a) Ergebnisse des BMBF-Verbundsvorhabens mechanisch-biologische Abfallbehandlung, MuA Lfg. 4/01 in Müll Handbuch, Band 5; Erich Schmidt Verlag; Berlin

Tabasaran, O., and Rettenberger, R. (1987) Grundlagen zur Planung von Entgasungsanlagen in MuA Lfg. 1/87 in Müll Handbuch, Band 4; Erich Schmidt Verlag; Berlin

Tchobanoglous, George; Theisen, Hilary; Vigil, Samuel A. (1993): Integrated solid Waste Management. Engineering principles and management issues. Mc Graw-Hill.

Teller, P., Denis, S., Renzoni, R. and Germain, A. (1999) Comparison between the incineration and the co-combustion in cement plants of industrial wastes using a life cycle approach, in 7th LCA Case Studies Symposium SETAC-Europe

Tsilemou K., Panagiotakopoulos D. (2004) Estimating Costs for Solid Waste Treatment Facilities. In: Proceedings of the ISWA World Environmental Congress and Exhibition, Rome, Italy, 17-21 October.

Tsilemou K., Panagiotakopoulos D. (2005) A Statistical Methodology for Generating Cost Functions for Solid Waste Treatment Facilities. In: Proceedings of the 5th International Exhibition and Conference on Environmental Technology, Athens, Greece, 3-6 February 2005 (in Greek).

Tzimas, E., Peteves, S. (2003) Controlling carbon emissions: the option of carbon sequestration. Office for Official Publications of the European Communities, Luxembourg.

UNECE (United Nations Economic Commission for Europe) (2004) http://w3.unece.org/stat/scriptsdb/showResults.asp.

UN-ESA (Population Division of the Department of Economic and Social Affairs of the United Nations Secretariat) (2003) World Population Prospects: The 2002 Revision and World Urbanization Prospects: The 2001 Revision, http://esa.un.org/unpp, September 2003

UN-Habitat, Global Urban Observatory (2004); http://www.unhabitat.org/habrdd/statprog.htm, April 2004

Vogt, R., Knappe, F., Giegrich, J., and Detzel, A. (2002) Ökobilanz Bioabfallverwertung, Untersuchungen zur Umweltverträglichkeit von Systemen zur Verwertung von biologisch-organischen Abfällen. Erich Schmidt Verlag, Berlin, Germany.

Wallmann, R. and Fricke, K. (2002) Energiebilanz bei der Verwertung von Bio- und Grünabfällen und bei der mechanisch-biologischen Restabfallbehandlung. In: Loll, U. (Eds.), ATV Handbuch – Mechanische und biologische Verfahren der Abfallbehandlung. Ernst & Sohn Verlag für Architektur und technische Wissenschaften, GmbH, Berlin, Germany.

Waste Watch, 1999 What people think about waste? Waste Watch/NOP Research, London. http://www.wasteguide.org.uk/issues/mn_public_-mresearch_public.stm

WISARDTM (1) Ecobilan-PwC (2003) LCI du recyclage du papier/carton en papier journal.

WISARD™ (10) Ecobilan-PwC (2003) LCI du recyclage des briques alimentaires en ouate de cellulose.

WISARD™ (2) Ecobilan-PwC (2003) LCI du recyclage du papier/carton en carton ondulé.

WISARD™ (3) Ecobilan-PwC (2003) LCI du recyclage en four verrier du calcin ménager préparé.

WISARD™ (4) Ecobilan-PwC (2003) LCI du recyclage de l'acier issu des centres de tri en acier secondaire.

WISARD™ (5) Ecobilan-PwC (2003) LCI du recyclage des boites boisson en aluminium, en aluminium secondaire (boucle fermée dans le corps de boite).

WISARD™ (6) Ecobilan-PwC (2003) LCI du recyclage des corps creux PEhd en flacons multicouches à partir de granules.

WISARD™ (7) Ecobilan-PwC (2003) LCI du recyclage de paillettes de PET régénéré à chaud en préformes tri-couches.

WISARD™ (8) Ecobilan-PwC (2003) LCI du recyclage des films plastiques en sacs plastiques.

WISARD™ (9) Ecobilan-PwC (2003) LCI du recyclage de plastiques mélangés en piquets.

World Bank (2003) Global Economic Prospects

World Commission on Environment and Development (WCED) (1987) Gro Harlem Brundtland, Chair;. Our Common Future. Oxford University Press, Oxford

Zegwaard M J (2000) Leidraad GIHA – Gescheiden Inzameling Huishoudelijk Afval systematiek voor verbetering afvalscheiding. (Report by De Straat Milieu-adviseurs for Provincie Noord-Holland en Gewestelijke Afvalstoffendienst (GAD) Gooi- en Vechtstreek)

Zeschmar-Lahl, B. and Jager, J. (Eds.) (2000) Mechanisch-Biologische Abfallbehandlung in Europa, Parey Buchverlag Berlin (eds.) Berlin.

Zickiene, S., Ruskus J. (2001): Individualaus buitiniu atlieku tvarkymo modeliai: apklausos rastu duomenys, in Environmental research, engineering and management, 2001.No.4(18), P.19-29, Kaunas)

***ibidem*-Verlag**
Melchiorstr. 15
D-70439 Stuttgart
info@ibidem-verlag.de
www.ibidem-verlag.de
www.edition-noema.de
www.autorenbetreuung.de

www.ingramcontent.com/pod-product-compliance
Lightning Source LLC
Chambersburg PA
CBHW061137220326
41599CB00025B/4273
9783898215190